T0304978

WATER AND PEACE

DR ALAIN GACHET

WATER AND PEACE

A journey through the world's
most explosive conflict zones
in search of deep water

ARCADIA BOOKS
LONDON

First published in Great Britain in 2023 by Arcadia Books

An imprint of Quercus Editions Limited
Carmelite House
50 Victoria Embankment
London EC4Y 0DZ

An Hachette UK company

A CIP catalogue record for this book is available from the British Library.

ISBN (HB) 978 1 52942 688 5
ISBN (Ebook) 978 1 52942 689 2

10 9 8 7 6 5 4 3 2 1

Typeset by Jouve (UK), Milton Keynes
Printed and bound in Great Britain by Clays Ltd, Elcograf S.p.A.

Papers used by Quercus Books are from well-managed forests and other responsible sources.

CONTENTS

List of Illustrations vii
Introduction: Daktari Wa Maji – "The Water Doctor" 1
A Brief Geological History of the Earth and the Appearance of Water 11
1: I Started with Oil 17
2: Initiation with the Pygmy People of Gabon 29
3: Rough Shift in Kazakhstan 41
4: Back to Oil in Congo-Brazzaville 47
5: Lost Illusions 55
6: The Brooding Aura of Gaddafi 61
Water Is a Magic Molecule 69
7: Feet in the Mud, Head in the Stars 77
8: Between Pride and Light in Darfur 85
9: First Rewards in the United States 103
10: The Front Line in Darfur Is Expanding 107
11: Disillusion and Pain 115
12: The Fall of a Dangerous Man 121
13: Drought Impacting the Horn of Africa and Middle East, 2009–2011 131
14: Protected by a Mongolian Battalion in Chad 137
15: A Surprising Outcome in Ethiopia, 2011 159
16: Discovery in the Turkana Region 163
17: Kawergosk Refugee Camp in Kurdistan 189
18: The Mosul Dam 193
19: Chemya! Chemya! 199

20: A Night with the Survivors 209
21: Lost in Kurdistan 217
22: Major Challenges Lead to Major Opportunities 221
23: Ultimate Rewards 227
24: What Can We Hope For? 231
Acknowledgements 237
About the Author 239

LIST OF ILLUSTRATIONS

Clearing hacked out of the forest with chainsaws to allow a helicopter to drop seismic teams and their equipment. © Alain Gachet

Water vines in Gabon primeval forest, in the Lambaréné area: the healthiest way to quench your thirst. © Alain Gachet

The author waiting for the helicopter on a helipad in the Gabon forest with his Pygmy friends: machete-wielders, compass-readers and dynamiters. © Alain Gachet

Tough crossing through a deep swamp with the help of the Pygmies, trying to protect my bag and notes. © Alain Gachet

With my interpreter Iouri Kouchnariov in front of the Kazakh Council of Ministers on the very day of the coup d'état, 19 August 1991. © Alain Gachet

Barricades in front of the Soviet parliament on 22 August 1991. © Alain Gachet

Radar image of the great leak covering several thousand hectares of land in the Sirte Desert, Libya. (Interpreted radar image by RTI.) © RTI

First WATEX™ image produced in Darfur by RTI, between Chad and Sudan, compared to a Landsat optic image. The unknown world of underground water is revealed by yellow and blue trends, with its black holes and star clusters. © RTI

Animals drop dead of thirst all along the Chad border: they are all that remains of the worldly goods of the Darfur refugees. © Alain Gachet

Darfur refugee children in a sandstorm on the Chad/Sudan border, awaiting help, in July 2004. © Hélène Caux

The yellow imprint of my hand on a tree to indicate a drilling point in eastern Chad, July 2004. A sign of hope. © Alain Gachet

Landmines in the Lobito-Catumbela area of Angola. © Alain Gachet

Satellite image showing the Darfur area, the Sudanese refugee camps in Chad (Iriba, Abéché, Oure Cassoni) and the camps for displaced people in Sudan (blue triangles). © RTI

Young women at the well in West Darfur, July 2004. © Alain Gachet

Pygmy prospectors pouring sandy water into a sluice box to wash and extract gold. © Alain Gachet

The Ennedi plateau between Iriba and the Fada oasis, taken from my plane window. © Alain Gachet

Prospecting at the base of the high Ennedi cliffs, under the protection of the MINURCAT. © Alain Gachet

Negotiating in front of the APC tank. L to r: Alain Gachet, Haroun Tagabo, Mongolian Major Ba'Atar. © Alain Gachet

Mongolian soldiers from the MINURCAT preparing the barbecue. © Alain Gachet

Strategy in the tent of the MINURCAT Mongolian batallion at the Haouach camp. L to r: prefect Issaka Hassan Jogoï, Alain Gachet, Mongolian Major Ba'Atar and Tunisian captain and UN observer Mamdouh. © Alain Gachet

The author working on geological prospection escorted by Gansukh, the Mongolian aide-de-camp, who was also the unarmed UN observer along the Ouadi Haouach. © Alain Gachet

A young Goran girl from the Toubou people, riding her camel to pull the rope and lift water from the well to water the herds. © Alain Gachet

Turkana orphan digging into the dried-up riverbed, looking for water for her two goats. © Alain Gachet

WATEX™ image of Turkana revealing five giant aquifers (yellow polygons) in the black holes of the underground galaxy: Lotikipi, Gatome, Nakalale, Kachoda and the Lodwar Lokichar Basins. © RTI

Drilling team cooling off in a pond of drilling mud at the Lotikipi 1 exploration well. © Edwyn Adenya

The first exploration well in Lotikipi, drilled down to a depth of 330 metres. © Sylvie Boulloud

Children dancing round a well in the Lodwar aquifer, discovered by RTI and drilled in July 2014. They had never before seen that much water. © Edwyn Adenya

Children in Turkana, in the lush vegetable plantation made possible by the new wells. © Alain Gachet

High dike on the Mosul Dam that stands at 113 metres above the downstream bed of the Tigris, with the hydroelectric power plant in the foreground. © Alain Gachet

The author overlooking the Kawergosk camp housing refugees from Kobani in Syria, and from Bakhdida near Mosul in Iraq, in January 2015. A young Kurdish soldier played the mandolin while ten kilometres away the bombing started near Mosul. © Alain Gachet

Nawzad Hadi Mawlood, the governor of Erbil, who gave me the passes required for the evaluation missions in zones that had been recovered by the peshmerga. © Alain Gachet

Rabia hospital bombed out by the coalition forces, especially British fighter bombers. © Alain Gachet

Military control by peshmerga on the road to Sinjar from Rojava. © Alain Gachet

Satellite map showing the itinerary (blue line) taken by the author in the February 2016 mission. Aquifers are inside the blue polygons. The southern aquifer was occupied by Daesh on the front line and mined, as were the aquifers to the north of the sacred Yazidi mountain. © RTI

Sheikh Zakho, spiritual leader of the Yazidi community, mourns the extermination of his entire family, buried alive at the base of the mountain by Daesh. © Alain Gachet

Leader of the group of Yazidi fighters, a patriarch now driven by hatred. He is still strong enough to lead the resistance movement in the mountains with his one remaining son. © Alain Gachet

Torched and blasted vehicle at the entrance to the town of Sinjar. © Alain Gachet

Sign informs Daesh vandals, "Yazidis live here".

Russian sign daubed on storefront by Chechen Islamist fighters.

Surprisingly, the impressive grain silos of Sinjar still tower over the plain, with slight scars at the top. The Sinjar plains were famous for their wheat crop. © Alain Gachet

The Yazidi Peacock Angel "Melek Tawûse" on an armoured vehicle captured from the enemy on the front line, facing the village of Ranbusi controlled by Daesh. © Alain Gachet

On death road at the southern foothills of Djebel Sinjar, women and children were stripped of their clothing before being put to death or sold into slavery. February 2016. © Alain Gachet

L to r: Astronaut Leroy Chiao, Gwynne Shotwell (president and CEO of SpaceX), the awardee Alain Gachet, and Daniel Lockney, the Technology Transfer Program executive at NASA Headquarters in Washington, D.C. at the time. © Alain Gachet

Mala Bakhtiar (left) from the Politburo, executive secretary of the Patriotic Union of Kurdistan (PUK), whom I met in Sulaymaniyah on 24 February 2016, with journalist and fixer Bakhtyar Haddad (right), killed in Mosul on 19 June 2017. We listened to his perspectives on the PUK. © Alain Gachet

Water Stress by Country in 2040 (World Resources Institute report).

LIST OF ILLUSTRATIONS

Monster dust storm on the tarmac at Niamey, Niger, June 2020. © Jean Michel Clair

Map of RTI worldwide operations cited in this book.

East African rift fracture lines stretch more than 7,000 kilometres from South Africa to the Middle East.

INTRODUCTION: DAKTARI WA MAJI – "THE WATER DOCTOR"

I am not a doctor of men but a Water Doctor: I locate water in the most desperate areas of the world, to heal the wounds that man has inflicted upon himself and his environment.

The discovery of deep water under the vast stretches of arid lands of Africa and the Middle East has been my main focus over the past twenty years of my career. I continue to pursue it, despite the challenges that it offers in a turbulent world, full of violence and uncertainty and undergoing major geopolitical and climatic shifts. The reason is human, and humane. Water discovery is not a job – it is a vital lifeline.

Water is a miraculous substance; a scarcity of it has, as we know, a devastating impact.

According to the last statements of the US administration in June 2022, "more than two billion people today lack access to safely managed drinking water, and nearly half the world's population lacks access to safely managed sanitation services. An accelerating climate crisis will increase pressure on water resources during this decade, and half the world's population is expected to face severe water stress by 2030. Water insecurity threatens lives and livelihoods and can lead to popular unrest and population movements. These conditions can undermine states and governments – particularly those that are already experiencing weak state capacity or legitimacy".

*

My initiation into what I am doing now took a very long time. For as long as I can remember I have been pursuing a childhood dream of exploring what lies beneath the Earth's surface: treasures of gemstones and minerals, inspired by Heinrich Schliemann, who discovered the treasure of King Priam in Troy. The age of reason led me to explore for hydrocarbons in an international oil company, from the steppes of Kazakhstan through the deserts of Arabia down to the jungles of Gabon. It is only for the last twenty years that I have been focusing on exploring this elusive, vital resource: water.

I write this as I sit beside the sea on the French Riviera in Sainte-Maxime – directly across the harbour, I can see Saint-Tropez. It's a luxurious hive of activity in August, bursting with every variety of designer brand, and frequented by the rich, the famous, and tourists who want a slice of the action. Outside every designer shop, a queue of people snakes down the street, all of them keen to get hold of something with a logo on it; expensive sports cars zoom around the streets, which are lined with smart restaurants serving haute cuisine. Docked in the harbour are the most extraordinary array of enormous yachts, each worth millions of euro. It is the beating heart of the privileged, wealthy elite. The sun is shining; people are on holiday; the mood is one of relaxation, luxury and contentment. And yet, when I turn my head to the mountains nestled against the harbour on my left, there's an extraordinary sight. A wildfire – apparently the result of a discarded cigarette on dry, hot land – has been raging, the sky a beautiful, eerie pink, which at first looks rather like an effect of the sun, until it becomes apparent that it's a reflection of the fierce flames blazing out of control. A steady stream of fire engines roars past on the road in front of the house – a seemingly never-ending line of red beetle-like trucks in convoy, for hour after hour, heading in the direction of the blaze, a constant supply of them, it seems. But even more extraordinary are the planes, bright yellow in colour, which fly in formation to the surface of the Mediterranean Sea, where they suck water into their enormous bellies and fly over the flames to dump it, circling back around to repeat the process over and over again. The distinctive colour of the planes is marked by dark ash and burns as the flames lick at them when they descend to release their loads. It's a heartbreaking sight, but an extraordinary one, too – all around us is the evidence that we are destroying our planet.

An interest in exploring the world around me came from an early age. I was born in Madagascar and had, as a child, a sort of daydream about my origins, a way of making peace with what I perceived to be a distance, a lack of understanding, between myself and my siblings, my parents: I fantasised that I was in fact from a different planet entirely; that I crash-landed in the forests of Madagascar on a reconnaissance visit, and was rescued by a man hunting for honey, nestled where I'd landed in the treetops. My saviour then took me to my "adopted" father, the guardian of the forest, who taught me about the natural world around me. In fact, my father – real, of course, not adopted – was a botanist, and a forest engineer, and through him my fascination with the natural world around me began. And of course, though my fantasy was a childish whim, there's a truth in it, too, something else that has always been a source of fascination to me: we are all of us – in a sense – children of the stars, physical manifestations of the result of the Big Bang, the evolutionary result of what happens to dust. I have always had a deep interest in space and exploration in tandem with a love of the natural world and the jungle – my childhood allowed me to explore both.

Physics is a fantastic tool; it renders that which is unexplainable – the unfathomable nature of the universe, the sheer expanse and mystery of it – into something that's tangible, that has an answer, an equation, like the Fibonacci series, which helps us to create a sort of mathematical architecture of the universe. Strict agnosticism or atheism has always struck me as rather naive philosophy – for there's a sense in which the nature of God is revealed by the beauty of nature; spirituality is our way of engaging with the mysteries of the universe. There's daunting beauty and harmony in a rainforest, as I observed as a child, and a harsh brutality and pain around its destruction – millions of years of balance and harmony, destroyed in seconds with a chainsaw or a wildfire.

But sometimes it's an urgent need for survival, a need for an immediate answer to an emergency situation, that creates a sort of destruction. I once encountered a Congolese refugee, who was deliberately burning down great areas of forest and collecting the dead animals and birds that fell from the trees in order to feed his family. Our growing populations, too, are creating an imbalance; 100,000 square kilometres of forest are burned down every year, and it's accelerating.

After my childhood in Madagascar I left for Paris, and studied quantum mechanics and nuclear physics; after university I went directly to mining school, where I continued my fascination with the origins of space and the mystery of what lies beneath our feet in equal measure. Suddenly I was studying the genesis of the Earth's crust, continental drift theories, the magnetism of the Earth, and all of the universe's properties – including, of course, water, and how it appears on Earth. I quickly turned to geology and geophysics, and to the exploration of the invisible underground world.

Various teachers influenced me along the way – most particularly, Professor Bernard Diu at University of Paris VII, himself a student of Richard Feynman* and colleague of Claude Cohen-Tannoudji†, both awarded the Nobel Prize in Physics. Through them, I grew to love and understand quantum mechanics and nuclear physics, for they opened the sacred doors to the universe and took me from the infinitely small to the infinitely large.

They generously provided the keys at a critical time of my life as I became a mining engineer, an explorer, a discoverer, and then an inventor. I also consider myself an archaeological explorer, continuing to discover ancient kingdoms in Africa and vanished treasures in the deserts in the Middle East and the jungles of Central America.

However, what has marked my life most deeply in the past seventeen years is the invention of a new process for surveying for

* Richard Phillips Feynman (11 May 1918–15 February 1988) was one of the most influential physicists of the second half of the twentieth century, due in particular to his work on superfluid helium, quarks and relative quantum electrodynamics.

† Claude Cohen-Tannoudji is a French physicist born in Constantine, Algeria on 1 April 1933. He was awarded the 1997 Nobel Prize in Physics for his work on laser cooling and confinement of atoms.

water. This work has enabled me to pursue a "Great Dialogue" with nature. I know of no description or words that could begin to capture the utter depth of the grandeur and the beauty of this world.

This in part explains my optimism even in the face of tragedies I have witnessed.

The supercomputers of the past, which gave birth to the Internet, have become common domestic and business equipment, alongside the latest mobile-phone technology.

The list of what they can do is long and far from exhaustive, and today we use these technological marvels without even realising it, taking them totally for granted. They are everywhere you look. Some of these tools have allowed us to explore the very generosity and beauty of Planet Earth.

As a young man I was fascinated by NASA, in particular the period that today is referred to as the Space Age. This passion has been a constant throughout my life; the conquest of outer space influenced my earliest memories.

Without NASA and all the images NASA has provided free of charge to the entire world via the Internet, my new profession in deep groundwater exploration would never have come about.

My work has brought me to undertake mineral exploration in primeval forest side by side with Pygmies, who have taught me a great deal. Pygmies initiated me into the world, as I like to recall when speaking to audiences at international conferences.

This often brings about a chuckle when I mention it in academic contexts, but it's a relationship that's proved vital to my work: from the Pygmies I learned ancestral knowledge as we walked and talked in the forests. I have cross-referenced their intimate, detailed knowledge of the land – land that they study, which provides them with all of their earthly needs – with my radar images. Pygmies helped me discover previously unknown mineralised geological structures. Indeed, that is how I discovered my first gold mines and iron mines in the heart of the Central African jungle.

The marriage of modern technology and ancient knowledge is an important one – myths and ancient stories are basically transposed forms of knowledge that hold often-unsuspected riches for us. Homer's *Iliad* and the *Odyssey* enabled Heinrich Schliemann to rediscover the cities of Troy and Mycenae; sections from the Old Testament guided the archaeological digs in Mesopotamia that I would embark upon with the French Archaeological Survey in Iraq, during breaks between two oil-drilling operations in the North Sea.

In 2002, I was using radar to prospect for oil on a mission for Shell when I noticed that the radar was showing enormous water leakages in the Sirte Desert in Libya.

With radar frequencies, I suddenly understood that what I was seeing was in fact several billion cubic metres of water being lost from Muammar Gaddafi's so-called "Great Man-Made River". This leakage had never been detected before. Identifying this leak was the starting point for my invention, the WATEX™ (for "Water Exploration") system, an algorithm that I worked on unceasingly for the two years that followed. It is an expert system that eliminates distracting surface effects, allowing you to see only what you're focusing on – it's akin to the Hubble telescope, which removes the distortion of atmospheric effects.

It enabled me to rapidly discover spectacular deep aquifers, amazing sources of water. My driving ambition could now be turned to implementing this groundwater algorithm.

I've had many triumphs, and some disappointments, along the way: the discovery of an aquifer underneath a dehydrated desert should, of course, be a cause for celebration and jubilation, but this is not always the case. Political leaders, when facing drought and hardship, often focus on "the problem" without asking themselves the right questions about what technological solutions can help.

Politicians are generally not scientists, after all; the focus becomes one of politics and not pragmatism. Selfish interests also significantly influence behaviour – country leaders are keen to be re-elected, and often face serious ethnic divisions. These governments are then reluctant to help people who are not supportive of their cause.

And of course, crucially, there's the delicate topic of transboundary aquifers – these vast underwater aquifers don't conveniently situate themselves within the arbitrary land boundaries of a particular country. Water has no passport – in the modern world, this is unacceptable to us, and it generates a sense that "the neighbours are stealing what is ours". Water can create and maintain wars and conflicts for generations; geopolitical destabilisation is often the result of this sort of squabble. Moreover, with deep water we are facing a serious change of paradigm.

For surface water, lakes and rivers, whoever controls the upstream controls the downstream, as General de Gaulle once said, and this is exactly what has happened until now with Turkey controlling the course of the Tigris and Euphrates rivers, which has had a dangerous impact on Iraq and Syria downstream.

For deep underground water, it is radically different. Whoever controls the downstream controls the upstream; if you tap water down a water tower, you will deplete it.

If we don't find a solution – technical, economic, political – the groundwater challenge of this century will have terrible consequences, including civil war, massive migrations and the destabilisation of states, and this context is exacerbated by the consequences of climate deregulation.

Today, we are working on several important groundwater exploration projects on behalf of several governments and financial institutions who have called upon our services. In this chaotic and mutating world it is still difficult to establish sound predictions, but one thing is certain: there is an increasing need for water in a world facing the multiple consequences of climate change and population growth.

Alain Gachet, July 2022

A BRIEF GEOLOGICAL HISTORY OF THE EARTH AND THE APPEARANCE OF WATER

If you are interested in the Earth, it is important to know its intimate character. The history of the Earth starts some 4.5 billion years ago, just after our planet took shape. At that time, the Earth was very different from the world as we know it today. There were no oceans. There was an atmosphere saturated with poisonous gases. Like all the other planets in the solar system, Earth was a hot aggregate of molten space debris, intensely bombarded by asteroids, which increased its mass. Earth was a dark planet with a burning core of iron and nickel that was enveloped by erupting molten rock undergoing nuclear reactions and exhaling a wide pattern of toxic gases. All the elements then segregated into different layers as a function of their density, which is why today Planet Earth has a dense core of iron and nickel surrounded by lighter elements of aluminium and magnesium.

Water, the most common material in space, existed even before the sun. Water was already present in space in the form of ice. From the very beginning of the constitution of the Earth, cosmic water was one of the basic elements of the Earth's crust, in combination with the other crustal elements. This cosmic water was a free element, in the form of vapour, accompanied by nitrogen and methane and heavily charged with carbon dioxide. Part of this cosmic water was contained in the magma.

Between 4.6 billion years and 3.9 billion years before our era, Earth was again subjected to intense asteroid bombardment. It is also possible that comets with cores of ice fell to Earth and contributed to adding more water vapour to Earth's atmosphere at least 3.8 billion years ago.

These events, after a slow cooling phase, gave birth to the primitive sea, which contained all the ingredients that gave rise to the very first traces of organic life.

Blue algae known as cyanobacteria lived on the nitrogen cycle and released a poison: free oxygen. They produced the earliest geological formations, which we call stromatolites. Imagine a primitive ocean without continents, with submarine basaltic iron-rich volcanoes exhaling heavy metals through black smoking undersea pipes, feeding one variety of bacteria that had no competition.

These bacteria, with exclusive access to feed on the nitrogen cycle, began to lay down a vast and thick layer of primitive algae. Their dead bodies, which coated the ocean floor over the course of several billion years, generated an extraordinarily thick mattress of sediments on the volcanic and hot sea floor. At this point, these sediments began their metamorphosis under pressure and temperature.

During their life cycle, these algae released the poisonous oxygen molecule in iron-saturated waters. It was at this time that trillions of tons of dissolved iron captured the oxygen and precipitated as solid iron oxide, forming the first iron-rich sedimentary rock on top of the cooled basaltic magma crust. This process gave birth to the richest iron deposits on Earth today, in Brazil, Australia, South Africa and Mauritania. They are called banded iron formations, or BIF.

I discovered all these wonders a few years ago in the Drakensberg Mountains while exploring the remote western parts of Limpopo, a north-eastern province of South Africa. At first glance, I did not understand the nature of these huge vertical cliffs, but then, looking closer, I discovered their horizontal brown iron beddings interlaced with white layers of accumulated algaes: the geological map of the area told me they were almost three billion years old. It took me more time to understand what had happened.

As the Earth gradually cooled, residual internal heat created a bubbling at the surface, triggering with the Earth's rotation the mechanism for future continental drift. At that time the land masses, exclusively made from light-metamorphised bacteria beds, started to float, like a crust that grew thicker over time, merge, then split and collide on the surface of molten basaltic magma. They were nothing like the continents we know today.

The gaseous atmosphere full of water vapour, nitrogen and carbon dioxide protected the surface from the sun's UV rays and enhanced cooling. As the cooling continued, the water vapour condensed and formed a thick layer of clouds.

Our planet had sufficient mass to retain its gases and form an atmosphere. It is not by chance that water appeared on Earth and settled here. Earth, as we know, is the perfect distance from the sun, neither too close nor too far away. Much of its water changed from a gaseous state to a liquid state, and flowed abundantly on the surface of the newly formed crust that emerged from the primitive ocean. With the emergence of these new land masses and oceans, everything was in place for life to emerge on the "blue planet". The oceans were formed and the Earth's continents began to come together – plate tectonic activity was underway.

Oxygen made its appearance very slowly, very gradually, from the activity of the blue algae rejecting oxygen as a poisonous gas, a poison immediately captured by the iron dissolved in the water, not leaving any chance for oxygen to increase in the atmosphere.

The drastic change in this atmosphere composition occurred with the appearance of the chlorophyll-based photosynthesis during the Cambrian revolution.

This magic chlorophyll molecule could break apart the strong carbon dioxide molecule to build brand-new organisms with a carbon-based architecture; rejecting the oxygen had begun to enrich the air composition half a billion years ago.

The current composition of the Earth's atmosphere is broadly 21 per cent oxygen and 78 per cent nitrogen. This proportion reflects the slow evolution of the atmosphere composition in the last 500 million years.

Chlorophyll accelerated changes to the atmosphere during the Cambrian era, leading to the Cambrian explosion of the Earth's biodiversity. New algaes, green in colour (derived from chlorophyll), no longer relied on nitrogen. They spread rapidly across all of the seas, capturing the abundant carbon dioxide to extract the carbon they needed in order to grow, while releasing oxygen into the atmosphere and the water. These green algaes gave rise to the huge diversification of plant and animal marine organisms that slowly began to inhabit the emerged land masses.

All these chemical and biochemical effects, combined with erosion of exposed land masses as the continents drifted, contributed to the creation of new sedimentary rock, which piled up in thick layers at the bottom of the seas.

This sedimentary rock then underwent further transformation as pressure and temperatures increased. It became more compact and harder, and gave birth to metamorphic rock such as marble and schists and then, even later, gneiss and granite, which made up the crystalline basement. These new rocks added layers to the primitive crust of bacteria sediments. Today, this crust is commonly referred to by scientists as Pangea. As the continents continued to drift, upheaval caused these underwater sedimentary masses to emerge and form mountain chains that exposed complex layers of fractured rock, leached and eroded by sand, wind and rain. Rainwater began to circulate deep inside these heterogeneous, emerged rock formations. And this is the point in the water cycle at which a hydrogeologist's work takes place.

1: I STARTED WITH OIL

After this brief description of the fascinating history of the Earth, you can understand why I joined an oil exploration company in 1977. I worked for twenty years for Elf Aquitaine, the French national oil company, itching to begin to explore the planet and to put my years of geoscience training into practice. It started with two years of purgatory, locked indoors in a research centre at Chambourcy, near Paris, doing research into seismic signal processing, a kind of mathematical processing that involved the first Cray-1 supercomputers we had the opportunity to acquire, thanks to our American partners at the Department of Defense. It was, forty years ago, tricky and painstaking work – the results of my observations and calculations were transcribed onto punch cards, which would occasionally get mixed up, resulting in the laborious process having to be repeated for days to have a chance to make it work.

A handful of frustrating years later, and after much pushing for a change, my work exploring the North Sea finally began – it was a huge relief. It was also, however, rife with other kinds of difficulties. We struggled against the North Sea elements, constantly subjected to violent weather, and we regularly faced new and unknown technical challenges. We were introduced to a newly invented offshore drilling technique that could be completed in extreme conditions – it was crucial to Europe's energy independence, and the challenge was daunting.

Before we could set off by helicopter to join the rigs, we had to pass an excruciating training tactic called, rather worryingly, the "trunk test" – claustrophobic and intensely unpleasant, this is a survival course that simulates a helicopter crash in the sea, anticipating that you are "lucky" enough to survive the initial collision and have instead found yourself bobbing on the water, trapped in the cockpit of a helicopter as it slowly sinks and becomes immersed in the North Sea's cold water.

The drill took place in a red-painted cage in water kept at 4° Celsius. There were several of us in a cramped space. The idea was that we maintained our calm, held our breath and calmly exited the cage. Easier said than done. Of course, the idea of it created anxiety before the simulation had even begun, and so we entered the exercise with quite a bit of trepidation. Once we were submerged in the unexpectedly capsized cage, we had to be careful to avoid panicking and knocking each other out in our scramble to escape; the sensation of suffocation and claustrophobia was palpable. The instructors ensured that no one would drown by keeping two deep-sea divers at the bottom of the pool, ready to rescue us if need be; but the dangers involved didn't stop the instructors from shoving us roughly into the water, sending us tumbling head over heels from the compartment. We repeated the drill regularly to hone the right reflexes. Anyone who panicked never made it to offshore operations;

and of course there are few things more likely to induce panic than, well, being told not to.

The oil platforms were located many miles out in the North Sea, near the Den Helder harbour north of The Hague, facing the island of Texel. We boarded helicopters, apprehensive of what awaited us, to be dropped on drilling rigs that were hung from steel pylons 30 metres above the water to avoid the strongest waves. The pilot had to land the helicopter regardless of the strong winds and fog, on a tiny helideck the size of a handkerchief, and our hearts pounded as the helicopter buffeted precariously from side to side as it came in to land. We were greeted by firemen who looked as though they were from outer space, clad in silvery fireproof uniforms, with pumps and fire extinguishers in their hands.

Once work on the rig began, hitting the deep underground reservoir of gas we had targeted a few thousand metres beneath the sea floor with the drill was another challenge, and we came to rely on, and have great respect for, the "Mud Engineers".

They either came from very good engineering schools, or were entirely self-taught on the rig; either way, they were masters at concocting mud cocktails of different densities that balanced the pressure of the gas rising up the drill string, suppressing fatal blowouts that could cause the loss of the rig and the entire crew. We were all aware of the risks, and we had to have confidence in the science of the mud engineers – combative and bad-tempered as they could be – penetrating the geological reservoirs that had held huge gas pockets for a million years.

This always drove adrenaline levels skyward. The rising gas had to be absolutely controlled, otherwise the sea surrounding the platform would bubble up like champagne and swallow any boats around in a huge explosion. In such cases, the heat is so intense that the sea can literally catch fire.

There's no easy escape, either, from a rig in trouble. Launching your lifeboat from 35 metres in the air into the sea would inevitably mean sinking straight to the bottom. With all the gas bubbles in the water, there would be no more suspension support for a ship: rising back to the surface would be impossible unless the boat was initially projected far enough from the oil rig and could leave the area with its own engine.

This may well be the phenomenon behind the disappearance of ships in the infamous Bermuda triangle. The bottom of certain ocean depths contains frozen gas in the form of methane clathrate (also called methane hydrate), stable at low temperatures and high pressure. Methane clathrate is a solid compound formed of ice and methane. These methane hydrates represent several million billion tons of gas on Earth, a veritable energy treasure but also a potential danger for mankind, from both a climate and a geological standpoint. In the event of seismic tremors, the methane can become gaseous and rise to the surface; any ship anywhere near this field of bubbles would suddenly sink.

On an oil rig, the other risk is a "blowout" – gas exploding as it comes out of deep-seated geological gas structures. Anyone reckless enough to smoke on deck would pretty much guarantee death for everyone, so safely rules onboard are strict, with tobacco as well as alcohol forbidden.

Twelve-hour shifts, day and night, for three-week periods, followed by the same periods of rotation back on land, are the norm for those who choose this line of work – it's hard work, often repetitive, in a difficult and potentially dangerous environment. Night and day, the platform vibrates with the rotation of the drill bit that grinds its way through the geological layers under the sea, centimetre by centimetre, layer by layer, for two to three thousand metres.

My job was to be in the drilling cab near the derrick, suspended 30 metres above the waves, summer and winter alike, and to determine what type of geological formation was in the process of being drilled. We knew we were getting closer to the gas by studying the type of microfossils being brought to the surface in the drill cuttings. As a geophysicist, I approved and interpreted the data, and made any corrections required to anticipate as accurately as possible the correct time of penetration in the gas reservoir.

Living conditions on the rig resembled those on missions to space, or on submarines.

We were far away from our families and completely alone, cut off from the world, especially during storms, at a time when mobile phones did not exist. We didn't really know each other, but conditions meant there was great solidarity between us.

Staff rotations took place daily, and every helicopter brought in a cargo of experts – Norwegian, Filipino, French, British,

Dutch, German. They were engineers, welders, cooks and work supervisors. An offshore rig is a Tower of Babel where countless languages are spoken, but English is the only one used professionally.

Our shifts took place day and night, on the different decks. We could only really talk to each other during meals, and as the kitchen was open – the refrigerators accessible at all times – mealtimes weren't formalised, and when we'd next encounter one another was unpredictable.

People became quite irritable after a few weeks, due to the alcohol and tobacco ban, so time on land inevitably included drinking sprees between Den Helder and Amsterdam; some never even made it as far as Amsterdam.

It was also during this period that I first felt palpable, genuine fear. We were flying in a helicopter off Den Helder, some eighty kilometres out to sea. I was feeling very optimistic and energised, as I'd defined a new structure for us to drill. The navigation instruments were the usual ones – altimeter, compass, gyroscope (GPS did not yet exist). Upon leaving the platform early in the morning, shortly after lifting off, we suddenly entered a freezing fog bank. The helicopter blades grew heavy with ice. The body iced up and took on more weight with each passing minute.

The helicopter was suddenly virtually imprisoned by this ice, and it was dragging us dangerously downwards towards the level of the sea, which we could not even distinguish in the fog. The pilot was sweating profusely and holding up his binoculars to peer to both port and starboard, inspecting the horizon with increasing desperation. Gusts of icy wind took over, shaking us in every direction. We appeared to be doomed. Suddenly, sheets of ice detached from the helicopter, and it was able to gain a little altitude. But the pilot had lost his bearings and the frozen controls were no longer responding. He let himself be guided by a radio signal, and we reached Den Helder with scarcely a drop of fuel left in the tank. I have to admit that the circumstances on that day forced my agnostic soul to pray.

I was confronted with another harsh experience in the middle of the Dutch winter, back on exploration in the North Sea. On Christmas night 1981 in the harbour of Den Helder, I boarded a seismic vessel called the *Navy Seal* to supervise an offshore geophysical survey. The rather crude American captain from Louisiana was hard work – he swore with ten words, exclusively about the indecent body parts of Jesus Christ, in a terrible soup of saucy sentences. We had taken off a bit too early, despite the bad weather and wild thrashing seas. As the boat tipped violently from side to side, we were all struck by seasickness – even the hardened seamen from Nova Scotia – and it was impossible to walk down the gangways, as violent sheets of seawater slapped at the prow. Fortunately, once we reached our work zone the weather had calmed.

A few days later I had finished our surveys and had to go back to The Hague for other projects while the ship continued on its route to Denmark. The company radioed a Dutch trawler to come and pick me up on the high seas and bring me to

Rotterdam. The old tub was used for bringing food and other comforts to alleviate the tough and dangerous tasks on the offshore rigs and was therefore empty upon its return, and so could take a lonely soul on board for the trip. My transfer took place at night, under floodlights; the big seismic vessel towered five metres over the upper deck of the trawler. There was no transfer net like in West Africa where, using a net and a crane, you were tossed from a small ship to the top deck of a platform 30 metres above it, like a sack of potatoes over the open mouths of the sharks waiting below. I was told that here, I was going to have to do the "Widow Maker", an expression I hadn't heard before ... and which was, due to the circumstances, rather accurate.

Initially puzzled, and terrified about what was to come, I quickly caught on to what it entailed. Two big arms stood ready to catapult me, without my meaningful consent, into the trawler five metres below us where three other burly men with (hopefully) even bigger arms were waiting to catch me. The two ships proceeded facing the swell, side by side, lights flooding the darkness. Huge salty waves splashed between the two hulls, drenching my trousers. The ship rose and fell with the swells.

We waited, drenched in water, for "the right moment" – it was impossible to imagine what that might be. Suddenly the two men in charge of the "throw" took advantage of a quick rise by the lighter trawler, which decreased the distance I would have to fall, and suddenly I was grabbed and flung from the *Navy Seal* into the air; the seconds seemed to stretch for ever, then the lower arms caught me. The manoeuvring had been timed to perfection, though in the time it took me to collect my breath and recover, I was very much counting my blessings. My baggage was tossed in just the same manner and I spent a restless night in the hold of the little fishing trawler, on a rather battered couch; I could imagine all too well what took place on it. The imaginings were not reassuring. Once back home in The Hague, I took a quick shower and, in the bathroom mirror, to my disgust I saw a pubic parasite crawling between my eyebrows.

Back to my office in The Hague: the seismic processing programs I had developed during my research work at the seismic centre in Pau, France, helped us to improve the result of our geophysics data. We knew that the company was taking great risks looking for gas under thick layers of salt, and our objective was to improve the rate of drilling success by finding the best possible subsea areas to drill; given the cost of a well, it was very important to avoid drilling useless structures. At the time we were working in the North Sea, a well ran to 3 million dollars (today exploration drilling can cost up to ten times that). Pinpointing the best possible areas limits the possible technical risks.

In the North Sea, the main obstacles that kept us from seeing the gas formations were huge salt domes that were several kilometres high. These domes had formed nearly 180 million years ago in the immense lagoon that separated America from Europe at the time. It was extremely difficult to identify the reservoirs of gas that lay at the base of these salt domes as they formed a kind of insulating liner – the experience was akin to trying to read the headlines of a newspaper through thick crushed glass. Any images we found of underlying gas were inevitably deformed by these salt domes. The solution was obvious, though a little complex: in order to accurately see the gas reservoirs, we had to first map the structure of the salt domes, even before taking radial seismic measurements, so as to more easily model the appearance of the gas reservoir located just below.

I played at being an optician, using very special corrective lenses to see through these domes. My bosses were sceptical – they wanted to impose a more traditional approach, which would have involved taking seismic soundings of several dozen kilometres in length along parallel lines. It didn't seem logical, given the problems we were facing – in my opinion we had to *acquire* lines, starting radially from the peak of the salt domes, to correctly map the image of the gas zone underneath.

The new technique worked: the first images that were acquired showed excellent, first-ever results. The new solution meant that the team of researchers who monitored our findings in the research centre in Pau took into account the corrections of the distortion caused by the salt. This work pinpointed a projected drilling cross-section that was highly accurate. We reached a gas reservoir 3,800 metres under the sea, within 20 metres of our calculations. This was an exceptionally good result when everyone had been expecting a resounding failure.

Along with the team of researchers from the research centre

in Pau, we received the Prix de l'Innovation, awarded by the directors of Elf Research in Paris, in 1985. Out of the 180,000 people who worked for Elf, only twenty or so were thus distinguished each year. I was given the title of technological innovator, for the discovery of gas in the North Sea. This prize earned me as many jealous enemies as it did friends . . .

Though we had been successful, I had unhappily acquired the label of "researcher" rather than "explorer", due to my design of seismic application models for complex geological structures; it was a title shift I didn't much care for, and the difference between "research" and "exploration" became paramount to my professional life.

2: INITIATION WITH THE PYGMY PEOPLE OF GABON

The success of our work seemed to incur the sort of wrath and envy sadly not uncommon in big companies; crossing the barrier between research and exploration was an offence punishable with professional banishment, and in July 1987 I was sent to Gabon for four years. I was frustrated by the sense that I had been demoted. The role consisted of spending day after day interpreting hundreds of kilometres of seismic soundings in a tiny office on the Port-Gentil peninsula. I had no real responsibilities, no longer took part in drilling decisions, and felt I had been sidelined.

The miserable situation began to take a turn for the better during my second year, when I was finally able to leave my closet-like office – it was when I first encountered the Pygmy people.

In July 1989, the head of operations came into my office and informed me that I was to pack my things immediately and leave for a seismic operation south of Lambaréné, which I was to supervise. This mission was in the heart of virgin forest next to Lake Onangué, near the famous Dr Schweitzer's mission on the banks of the Ogooué River. It was meant to be just one further vexation for me, but I took it as a positive windfall; after two years of desk work, I was finally getting out of the office and taking action in the field. We would have to travel by helicopter as the terrain was so dense; the few trails that existed were mainly used by gorillas and small, aggressive dwarf elephants called asalas, which in the thick vegetation could be spotted only at the very last minute. In this rather unforgiving landscape, there was the very real possibility of dangerous encounters with wildlife.

As our helicopter hovered above the camp on the lake, I was excited at the prospect that true exploration was finally beginning for me. Our tent camp stretched along a former airstrip on the lake. Our life there revolved around a sort of camp refectory near the bar, and the heliport itself, lined with helicopters. Every day, first thing, one of these helicopters would take us on our expeditions, deep into the forest: the radio room was the only thing that kept us in contact with far-distant civilisation.

At dawn, scarcely awake, we dragged ourselves into the first helicopter and were set down, after a half-hour flight over the monotonous green and dense rainforest, on a helipad that was little more than a simple clearing in the middle of a green mass of vegetation hacked out with chainsaws. The helicopter dropped us at the intersection of two seismic tracks, and immediately took off again. If weather conditions permitted, we would be picked up the next day, in the evening, in another clearing located at a different intersection of seismic recording trails.

Approaching primeval forest from the sky is an extraordinary experience. As you land, the rotor blows showers of branches, leaves and pebbles up into the air, swirling all around.

As soon as the helicopter touched the ground we would jump from the cab, amid the whirling blades, and we would shelter as quickly as we could from the projectiles the helicopter created as it lifted off.

In the aftermath of the helicopter's departure, total silence would fall over the forest canopy for a few seconds, before the jungle suddenly came back to life, with a cacophony of sound from strange insects, birds and monkeys: we would quickly gather our wits and find our bearings, and source the seismic tracks.*

Since two tracks come together at each helipad, it was important not to make a mistake, otherwise you risked finding yourself spending the night alone in the forest, a daunting prospect: as if it wasn't bad enough worrying about the gorillas and the panthers during the day, it was impossible to deal with the teeming fauna on the ground at nightfall. Spiders, scorpions, cockroaches, ants, snakes and centipedes guaranteed a sleepless night at best, not to mention the most dangerous enemies: aggressive mosquitoes, vectors of malaria and dengue.

A few days later, our work dictated that I went on a trip alone, and I found myself dropped from a helicopter with no one for company, listening to the sounds of the jungle around me, alone but completely surrounded by life. Two intersecting lines

* A seismic track is a narrow passage carved out of the forest with a machete, to install seismic sensors, known as geophones. Workers placing explosives also travel on seismic tracks.

awaited me, and I was to follow them, but it wasn't immediately clear in which direction. With the help of my compass I fortunately took the right way, but I had no idea that the forest track would be so much longer than I ever thought.

In a primeval forest, you find yourself walking with respect – it's like being in a monumental cathedral. You speak to yourself and with others only in whispers. The trunks of the great trees are like vast pillars of a cathedral; the leafy canopy that filters the rare rays of sun that reach the ground is like countless coloured stained-glass windows. The fragrant resin from the tall okoume trees is the scent of myrrh and incense. The entire majestic forest is a magnificent equilibrium of telluric, biological and climatological forces that has taken thousands if not millions of years to attain. The forest is the true miracle of evolution expressed in harmony, balance and beauty. It is a tangible manifestation of a kind of divine presence – in these moments of grace, there is a profound sense of peace, and of truth.

The banks of forest streams, scarred by iron and by fire, are flooded with sunlight. Tornadoes blow into these riverbed corridors and disrupt the ancient harmony, bringing chaos: riotous undergrowth, including parasitic vines that cling to centuries-old trees from the very roots to the highest crowns, is emptying these venerable trees of their sap and slowly killing them. In this secondary forest reigns the chaos of plant wars. It's almost impossible to penetrate this disorderly growth.

After a few minutes of striding on alone in this awe-inspiring, eerie landscape, I felt a presence. Someone was following me: there was no doubt about it. I could hear, somewhere behind me, footsteps, which stopped when I stopped. I called out in vain to the individual, who refused to come out. Their silence made my hair stand on end – whoever was tracking me, secretly, could only have negative intentions towards me. I was not armed for self-defence and my heart was beating uncontrollably. I started running to exorcise my fear; but to my horror, the steps resumed, faster and closer. Whoever, or whatever, was in pursuit, was matching my footsteps.

What danger was I facing? What person or animal had set their sights on me? I knew the forest was rife with panthers, but panthers only hunt at night. No one had informed us of any sightings of the infamous, highly aggressive asalas, and in any case the quick and supple steps that followed mine were not those of a heavy pachyderm. What was I doing all by myself in the middle of the hostile jungle? Why had I agreed to walk this track in the middle of the forest without a guide? Solely to not lose face among my oil peers? That was assuredly the reason. In this trade you never give up on a mission, especially a mission that is fraught with danger. It is often said that petroleum engineering is a macho trade (although this did not and does not preclude women from being very proficient in this line of work). But this was not the time to be thinking of such things. The steps were closing in and by now they sounded dangerously close.

Exhausted, defeated, sweating profusely, and submerged with fear, and worst of all clueless as to how I was about to die,

I found a big stick to give myself a little courage. Now all I had to do was to claw back a bit of dignity as I died and fight to the finish, even if I had only a paltry weapon. My shirt stuck to my back, and countless tiny black flies were drawn by my salty perspiration and flew into my eyes and nose. The flies made me pause; all of a sudden, the presence of them – they're known as "gorilla flies" – gave me a possible clue about what might be behind me. And indeed, through the foliage, I finally spotted it. An enormous gorilla was following me on a parallel trail, just a few yards away. Strangely enough, though it was terrifying in its size and I knew all too well how dangerous gorillas could be, it calmed me greatly to know the danger had been identified. At least I had an answer, a known opponent. I began to resign myself to my fate, given the imbalance in our respective strengths. I drew a deep breath, and calmed myself; perhaps the gorilla was simply curious about my intrusion. After all, I surmised optimistically, if he'd wanted to attack me, he would have done so at the start, when we both started down the track.

I assumed an air of confidence, almost indifference; it was my last chance, my joker card. I virtually whistled, and walked as normally as possible. I pretended to not even notice the gorilla. We continued down the track for what seemed like an eternity but was most probably no more than a few minutes. My tactic worked, to my intense relief, for my fellow traveller vanished as if by magic, and with it evaporated my fear.

Breathless after two hours of walking, I found the entire

seismic crew at the end of the track. Boris was the *boussolier*, the compass guy – he was armed with a string box, a device from which he could measure distance by means of an anchored silk filament wrapped around a wheel that revolved as he walked. He was to trace the straightest direction he could between the trees, and we were to follow it. There was Athanase, the machete-wielder, who was to carve out with his machete the straight path Boris traced for us. Once the track was open, the dynamiters, under the direction of Oscar Mavoumbou, a Bantu from Lambaréné nicknamed "Oscar Wild from the Bush", were to go through it with a hand drill to thrust the sticks of dynamite into the ground. The procession ended with Victor, the *boutefeu*, equipped with a magic box; he connected the sticks of dynamite to red and blue wires, and set off the blast.

I stayed behind with Victor, not to supervise his work and ensure he didn't mix up the wires, but because I was still catching my breath. I left Boris, Athanase and Oscar Wild's crew to continue opening the track. Victor clearly saw from the fact that I was hot and out of breath that I had just been through a challenge of some kind. I told him about the gorilla and he assured me that "Monsieur" would never have hurt me because "Mrs Gorilla and the baby gorillas" were not with him today. He spoke with a reverential, respectful attitude, as if he was speaking about members of his own family, describing the details of their daily goings-on. He quickly added that yesterday it had been important not to anger the male gorilla, because yesterday: "Monsieur Gorilla blocked the road, and the French topographer nevertheless wanted to force his way, and Monsieur Gorilla became angry, because his missus was crossing the track with the little ones." The gorilla was irritated and broke the topographer's sight device, leaving the mark of his teeth in the theodolite metal case – he was lucky; the gorilla could easily have ripped the topographer's arms and ears off. The fortunate topographer ran screaming away from the site and that very day took the first helicopter straight back to Lambaréné, then back to France. Gabon, he'd decided, was not for him. This amused Victor greatly and he laughed all the harder: "Monsieur Gorilla is really the strongest, not the white French topographer." Then he added mischievously, "The elephants are the best compass-readers in the forest, and they do not need, like white man does, any string box to lay out their path in the forest!" I later learned that the engineers who had laid out the route for the Congo–Ocean Railway (the COR) in the 1930s indeed followed the corridors created by the elephants tramping along the ridges of the impenetrable Mayombe forest.

We were sitting comfortably on the ground in a slight hollow, waiting for the blast-off. Victor had somehow sensed the proximity of a little forest antelope and had been calling out to it, imitating its call by pinching his nose with a leaf between two incisors, emitting two short, deep sounds. After a few seconds of silence, the tiny animal was standing in front of us, as surprised as we were, until it realised the trickery and bounded off into the forest. Victor was proud to have demonstrated to me his knowledge and his talent.

The radio attached to the blast-off box finally barked the order to fire. Victor connected the wires and, with dignity, turned the handle. Boom! Clumps of earth sprayed upward and outward all along the track. Mission accomplished. We could start down the path, behind Boris and Athanase, on the elephant trail; aggressive elephant flies and other bugs collected on my collar as we walked.

We spent the entire afternoon together; Victor kept up a constant patter on all the great trees in the forest, replete with images and names: azingo, ozigo, ozouga, izombẹ́, padouk, niangon, douka, iroko, limbo, moabi, zingana, dibetou, an entire repertory of hardwoods, plus the names of the genies that inhabited each tree, living in the leaves, flowers or roots. Some of these genies enabled post-menopausal women to continue to breastfeed their grandchildren, others could "give a man an erection as big as an okapi's"!

In that forest were all the sources of the pharmacopoeia used by the Pygmy world; it was a laboratory in and of itself, bursting with cures for every illness and ailment. The Pygmies are expert croppers of the natural world around them, relying on nature for solutions. To my chagrin, as neither a chemist nor botanist, a lot of the intricacies, the "recipes", were lost on me.

Just as we were about to cross a swamp, Victor showed me how to brush aside crocodiles by priming a stick of dynamite before tossing it into the water. A young crocodile was spewed out of the backwaters, with a scream akin to a child's cry. We cleared the swamp, a real obstacle course, and got covered in mud from top to bottom; there was something ludicrous, something hysterical, about the experience, and both of us were laughing like schoolboys.

We then had to take a dugout canoe that stood ready for us at the end of the track. It took us to the nearby dynamite storeroom in a remote barrack, to get a new load of explosives for the following days. Bakar Mafou, the warehouse guard, had been informed by radio that we were coming, and was expecting us. Victor, by now my unofficial guide and commentator, told me that Muslims were the best guards for dynamite because they did not touch alcohol. A few months earlier, he said, an explosive storage hall had blown up along with its tipsy Bantu animist guard.

Our canoe was filled with crates of explosives and it took us back to camp before night fell. The sun goes down fast at the equator, and by 6 p.m. it was pitch black. That night, the camp became another war zone – this time, it was us against the insects. Fortunately, the Pygmies knew exactly what to do. All around the big hut, which was open on four sides, they placed repellent torches they had made from the resin of okoume trees and from bark, which (we hoped) successfully repelled snakes, spiders and all the crawling insects of the jungle, at least for the night.

Crawling insects were thus dealt with; for flying insects – in fact, anything small, anything flying – their technique was not so great. We had to resign ourselves to being totally smoked out by the okoume-resin torches used to combat the mosquitoes, which, strangely enough, did not seem to bother our Pygmy friends one little bit. Tired of the struggle, I selfishly set up my own little mosquito net on a raised platform far from the creepy-crawlies, in a halo of acrid, stinky smoke.

There was no way you could leave the hut to heed nature's call, for the region was full of panthers, notorious night hunters. The Pygmies, normally fairly relaxed, were very cautious about our conduct once night had fallen. Men and animals showed one another mutual respect, and it was possible for them to live in harmony. But when it came to nocturnal hunters, there was no sense in becoming a tasty source of temptation . . .

During conversation that evening in the hut, Oscar Wild dared ask a question that had been nagging at him: "Say, Boss, gorillas are protected around here, and it's forbidden to eat them, is it not. But the Pygmies are not really protected, are they?" He saw from my indignant frown that his question brooked no response.

3: ROUGH SHIFT IN KAZAKHSTAN

These adventures did occupy my mind for a while, offering a release from the tedious desk work. But after being sidelined in Gabon for two years, I resigned from Elf and made plans to go and study for an MBA in the United States. My boss tried to make amends – he informed me that, following the fall of the Berlin Wall, all the Eastern European countries had opened up. The new president of Elf, Loïc le Floch-Prigent, was going to set up a commando squad of bad-tempered explorers, economists and negotiators to be the first to get new oil exploration licences in Russia and Kazakhstan, getting a head start over other big oil players, and he wanted me to be part of the team.

Six months later, I was at the base of the Tian Shan Mountains, the high range that borders the easternmost edge of the immense plains of Central Asia, near Lake Balkhash, facing Kazakh horsemen in a mosquito-free yurt. We began the first series of negotiations for oil exploration contracts with the newly formed Republic of Kazakhstan, emancipated by Gorbachev.

Kazakhstan, Brezhnev's sphere of influence, is located between Russia and China, and in the 1990s it appeared in no tourist guides. Reading stories from the campaign waged by Alexander the Great on the Amu Darya and Transoxiana seemed the only way of understanding this mysterious place, and these ancient tales gave me my first impressions of where and with whom I was going to be working. The Iron Curtain years of communism and the Semipalatinsk nuclear Polygon were among the game-changers. It was in Kazakhstan, over the course of thirty years, that the Russians had conducted their first military nuclear tests, above ground and with little concern for anything else.

My final mission in Kazakhstan, while leading a training seminar on petroleum economics for the Council of Ministers in Alma-Ata, ended in the failed coup d'état against Gorbachev on 19 August 1991 in Moscow.

That morning at daybreak we found ourselves surrounded by tanks driven up to the high walls of the Palace of the Kazakh Council of Ministers of Almaty. Devastated government officials whispered together, in front of a bank of telephones that were ringing off the hook. The situation was serious – they were criticised for economic rapprochement with the West, for trading their oil with capitalist businesses. They could be executed by firing squad after a hasty trial; in this chaotic context, I couldn't help thinking that my welfare was not anyone's uppermost concern.

After the initial fear and trepidation, I had a flash premonition. It was perhaps no more than an urge to reassure my new associates, and with them myself, but suddenly, I felt deeply persuaded that this new openness in the Soviet Union was inescapable, and that this putsch was like the final shudder of a doomed empire. My vision was clearly not shared by the men who had experienced the reality of the purges of the past seventy years. It was all they had known their entire lives! I assured them that our negotiations would be maintained and, if not executed as a spy, I would continue to work with them more quickly than they thought. I had no idea how I could be so sure, given all of my own apprehension about the dramatic, unforeseen circumstances in which we found ourselves, but my assumed certainty calmed the negotiation team.

Through repeated telephone calls with headquarters at Elf Aquitaine, we finally understood that real panic reigned in Paris, which is often the case when you are far from the context. Despite my desire to stay, I was obliged to cut this mission short, and I separated from my new Kazakh friends on the tarmac at the airport with a heavy heart. I had to jump into the very last Ilyushin aircraft leaving Almaty for Moscow, and hope to get one of the last flights on Air France from Moscow to Paris.

Getting to Domodedovo airport in eastern Moscow was no easy task. In the early morning hours, it was a rough ride trying to cross the bridges over the Neva River, guarded by tanks, soldiers standing at the top of the turrets, fingers on the triggers ready to fire their machine guns. My driver had the bright idea of having me hide under the back seat of the car while he wound his way amid the burned-out vehicles near the parliament, veering around barricades that had been set on fire, until he reached the Marco Polo Hotel in the centre of Moscow.

The employees at the hotel were in the lobby, in tears. This insurrection violated the new spirit of openness so desperately hoped for since the glimmerings of the perestroika (restructuring) sought by Gorbachev, who was now under house arrest. That night I tried to call my wife in France to reassure her that I was fine, and that she had nothing to worry about; she was six months pregnant and understandably anxious. Luckily, I got through quicker than expected and was able to talk things over calmly with her. All of a sudden, however, there was a deafening burst of automatic rifle fire somewhere nearby, which drowned out our conversation. My assertions that all was well suddenly sounded rather hollow.

Early the following morning my driver took me to the Elf offices in Moscow, just a few kilometres from the hotel. Tanks still barricaded the main roads, but adjacent side streets were open. At Elf headquarters, the secretaries were wringing their hands, their faces grim with apprehension at the prospect of banishment to the gulag. Deep-seated terror resurfaced, combined with a palpable sense of guilt for having worked for a capitalist business. Their recently found improved standard of living now appeared immoral in the face of the imminent danger of going back to hardline communism. They clearly remembered a time, just a few years ago, when collaborating with a Western firm – and an *oil* firm, to boot – would have been a political crime. I attempted to reassure them that this coup could not last, and that the new openness of their country was sustainable. Tears continued to flow, for with time we had come to know, respect and like each other.

The next morning, after a short debrief with the director of the subsidiary of my Alma-Ata mission, I was bombarded with phone calls from Elf Aquitaine in Paris, insisting that I leave Russia as quickly as possible. I tried to convince them that all the flights were full – I did not want to leave the country when the end of the coup was imminent, with the country right on the cusp of such a historic event, despite my fears for all of our safety. Suddenly, the walls began to shake: the glass rattled in the windows, the furniture began to jump, pictures fell to the floor, as though an earthquake was building. Outside the window there was a column of tanks travelling full speed up the main avenue towards the Russian parliament.

As I stood watching them, I had a phone call from my brother; he was watching CNN in a shop window in the French city of Valence in Rhône-Alpes. He informed me that tanks were rolling to support the new mayor of Moscow, Boris Yeltsin, who was at the parliament building, and to denounce the coup and call for a general strike.

It was an extraordinary moment. All of us from the office immediately streamed over to the parliament building a few minutes' walk away to see the Soviet flag come down and the Russian flag go up to the deafening sounds of an overjoyed crowd, clapping and whooping, even though smoke still poured from one wing of the building. On the balcony of the parliament building, Marek Halter represented France, alongside Boris Yeltsin as he addressed the crowds. That same evening, in front of the infamous Lubyanka prison where countless political prisoners had met tragic fates, the statue of Felix Dzerzhinsky – founder of the NKVD – was toppled.

By the end of this week, much earlier than I expected, the Kazakh team arrived in Moscow from Almaty. This event reinforced my confidence in my intuition.

What a warm welcome that was, after the week's improbable events, with a conclusion that we had scarcely dared hope for. Together we took a flight back to Paris to continue our negotiations. We were crowned with the sense of glory that inevitably surrounds those who have survived an international unpredictable disaster – it felt miraculous, surreal even. Strengthened by the experience, the negotiations that had begun in Kazakhstan for the Emba Basin oilfields south of the Urals continued up until 1992 for the oilfields in Russia, then in the Middle East, Syria and Qatar for the world's biggest gas deposits, which included the North Field.

Undoubtedly these successive operations in areas of upheaval paved my way for future work exploring water in every conflict zone on Earth. It was an amazing introduction, and I took away some reassurance about understanding the importance of relying on my flash premonitions in the face of seemingly insurmountable danger and uncertainty.

4: BACK TO OIL IN CONGO-BRAZZAVILLE

After this successful experience in Central Asia, I expected to be rewarded by being made director of Elf in Kazakhstan, and I was totally disillusioned when, in 1992, I was instead nominated as general secretary of Elf Congo in Congo-Brazzaville. The Berlin Wall had fallen, and with the end of the Cold War regional African geopolitics were destabilised and disoriented. Countries in the old blocs of the Non-Aligned Movement, under economic and institutional pressure, had to seriously review their strategies and their alliances with the Western world. It was a complicated time, and my role in Brazzaville in these murky waters would be more political than technical.

A democratic process was underway on the entire African continent, giving rise to strong opposition forces. Congo-Brazzaville was part of this movement and its historical leader, President Denis Sassou Nguesso, was approaching the delicate exercise of perestroika and glasnost (openness) with intelligence. A similar process was conducted on the other side of the river by President Mobutu, much more roughly; it ended in violence and in his eviction from power.

A long period of instability started in what was then Zaire, especially in South Kivu, in Katanga and also in Kinshasa, the capital, which stood just across the Congo River, facing Brazzaville, where I had settled with my family.

In a few months, in 1993, Angola and Congo-Kinshasa both sank into civil war. One of my first tasks, in the aftermath of the assassination of the French ambassador in Democratic Republic of the Congo (DRC), was to organise the arrival of Elf expats who were fleeing the city of Luanda on military cargos, and to help with their exfiltration to France; it was a time when mortar shells were lighting up the fronts of the major buildings of Kinshasa. We were only a river's width apart, and we heard the clamour and saw the fires and explosions mirrored in the river all night long.

Mass evacuation of Elf employees occurred simultaneously in Angola and DRC. High-capacity aircraft unloaded women and children from Luanda onto the airstrip in Brazzaville, all of them still reeling from the appalling scenes of fighting in the city centre. My wife and I gave them food and lodgings, and helped to organise their return to Paris.

The inauguration of the new democratically elected president, Pascal Lissouba, took place in the presidential palace of Brazzaville in September 1992 in the presence of all the foreign delegations. Three months later the militias were fighting among themselves in the very heart of Brazzaville, undoing the civil peace and the national unity promised by all the parties. Artillery began to roar on the hillsides and every day we picked up bullets and shrapnel that had hit the front walls of our house. My seven-year-old daughter could no longer attend school – it was too exposed to the shooting between the militia groups.

I was still front and centre for several reasons, among them the fact that in the absence of my director I represented the Elf

Congo firm, which had several oilfields in the Kouilou district and offshore in the Atlantic between Cabinda and Gabon. Maintaining their production guaranteed a certain economic stability and constituted a major strategic asset, both for income into the Congolese government and for France, whose energy security was also at stake. On the other hand, for the time being the second-largest city, Pointe-Noire, was calm, far from the disruption in the capital, and oil continued to flow freely to the Djeno terminal and refinery on the Atlantic coast.

With all the barricades and barrages in Brazzaville, my mission was first and foremost to ensure the protection of all Elf employees under my direction, who represented many different ethnic groups. It was a frightening task, though outwardly I attempted to be a calm, even paternal influence, accompanying my staff back to their homes amid the roadblocks and the intimidating militiamen. Some outlying districts were being pillaged and ransacked; in the centre of the city, we increasingly felt a net tightening around us.

I was often awoken in the middle of the night by calls for help. The wife of one of my colleagues belonging to the Lari tribe called in tears to inform me that her husband had been detained at a roadblock by vigilantes of the Bembé tribe. The next night it was the turn of the wife of a Bembé colleague – she was calling me to ask for my help in securing the release of her husband, who had been taken by the Mbochi tribe militia groups ransacking the town. I'd been watching the progression of each group, making notes of who was who, and so I had a way of being in contact with each leader. I called their chief and demanded the unconditional release of one of "my children". The message I gave them was clear: there was to be no negotiation. The man must simply be returned home safely, no questions asked. The chief released the man unharmed the next morning at dawn, to everyone's immense relief.

It presented an opportunity to attempt to reason with these factions. I took the reunited couple aside and, with my wife as witness, told them that if they ever captured any agent from any other ethnic group, their duty was to protect their prisoner and return them unharmed and intact with all their limbs – to afford them the same kindness that they themselves had just received. This watchful neutrality worked well throughout the entire conflict period, and no employee of Elf Congo in Brazzaville was killed or even hurt. But despite feeling I had in small part helped, and that I was doing the best I could in my role, I was beginning to feel stirrings of a vague disillusionment.

In April 1994, at the peak of the Congo crisis, the Rwandan tragedy happened. Vague rumours began to reach us in Brazzaville; soon, via the local and international press, we were hit with the horrifying truth of the systematic genocide undertaken by the Hutu regime under President Juvénal Habyarimana. Stories began to reach us first-hand via survivors who communicated with family members local to us. We heard from every clan, every group, every militia; each had a different story, a different perspective, a different take on the horror. Some accused the French of taking part in the genocide; others accused Belgian colonialism of creating the ideological foundations of such a conflict. We were truly afraid of regional contagion in Congo-Kinshasa and Congo-Brazzaville.

I was warned of a possible attack on the company's Falcon jet, which I flew on regularly between Pointe-Noire and Brazzaville. The Rwandan president Juvénal Habyarimana had just been killed in such a way, when his jet was shot down at the end of a runway with rocket launchers, sparking the Rwandan genocide.

One week later, I had the jet return urgently from Pointe-Noire to pick my two-year-old son and me out of the chaos reigning in Brazzaville. My wife had left Brazzaville a few days earlier with my seven-year-old daughter, who had had a total nervous collapse after witnessing armed confrontation in the city.

The pilot radioed me that given the threats around the landing strip, he would not go to the passengers' hangar but would remain, engines running, lights out, at the edge of the strip for immediate lift-off. The road was studded with barricades formed by tanks; we had to get through each obstacle in turn before reaching the airport at dusk. I lifted my son into the craft and jumped in behind him. We had not yet sat down, much less fastened our seatbelts, when the plane took off at full speed. All the lights were off, which enabled me to see the tracer ammo around the body of the plane. I only started breathing normally again when we reached cruising speed at high altitude, and I looked down at my sleeping son with gratitude and love.

A terrifying close encounter occurred one day as I travelled to the airport to fly to my family in Pointe-Noire. A barrage of rogue militiamen stopped me near the zoo – I had learned that, apparently, they tossed their enemies into a crocodile pit there. The chief mercenary, armed with a Kalashnikov, was paralytically drunk, and could have fired at my car at any time. To avoid a direct confrontation I stopped the car and got out, under a volley of insults and threats. I tried to catch the chief's gaze, but to no avail – his head nodded from left to right, and his bloodshot eyes stayed evasive. His hand clutched the weapon, and he smelled of blood; it occurred to me with dread that he had escaped an ambush, and was looking for a fresh kill at any cost. His desire was obviously intense.

I was not able to make eye contact, though I sensed that if I could, it would save me; a man who wants to kill cannot look you in the eye. The muzzle of his gun prodded my side. I remained immobile, paralysed with fear by what seemed inevitable – a summary execution, my body left in a ditch. The situation was so surreal, so horrifying, it became almost hollow; to my surprise, I felt incredibly calm. Important scenes from my life passed through my mind like a silent film running in slow motion. There was an immense sadness that spread from my heart into my gut, then coursed throughout my entire body like an unavoidable, implacable truth. It wasn't a fear of death itself; I'd concluded that there would be no pain in it, only oblivion. But the profound sadness was for my family. I would not raise my children. I would not be there for my wife, left to grow old alone without me. The sadness became a profound sorrow.

When all seemed lost, and death seemed inevitable, my walkie-talkie suddenly sprang to life. I reached for it as calmly as I could, and informed the officials expecting me at the plane that I had been stopped by vigilantes near the zoo, by the elephant statue, and that I was about to be shot.

I shut off the walkie-talkie and looked at the soldier of fortune, dazed with alcohol and weed. I told him in a robot-like voice, devoid of emotion, "Go ahead and kill me now, but" – I breathed deeply – "my friends will find you!" Just then his chief came out of the bushes and took him by the arm, muttering to me, "He wants to kill you, so get out of here, without a sound. Don't slam the car door, and drive slowly." I moved away from them slowly, my heart in my mouth, and drove without haste, gradually picking up speed. As soon as I knew I was clear of their sights, I raced to the airport.

The flood of adrenaline was overwhelming, and with it came a strong wave of hateful revenge; I wanted to go back, taking the airport guards with me, and slaughter the vigilante. Sense prevailed, however, and instead I jumped hotly into the plane, vowing that I would never again take that road without an escort.

5: LOST ILLUSIONS

In early 1994, Elf was privatised. Civil war continued to rage in Congo-Kinshasa and in Angola. I noticed unidentified aircraft with no registration lifting off from Brazzaville and Pointe-Noire. No one seemed to know who these planes were working for, but they flew with fake flight plans in the direction of Luanda or Huambo. Negotiators from Elf were involved in hellish goings-on between Angolese president José Eduardo Dos Santos, backed by the Soviet Union and Cuba, and Jonas Savimbi, his ferocious opponent, backed by anti-communist South Africa.

This cruel war, conducted with money extracted from underground riches, was being fought in the greatest ideological confusion possible. It left hundreds of thousands of mutilated victims. No one counted the dead; they were quickly buried and forgotten. There was a seemingly endless supply of young people from the teeming neighbourhoods who could go and clear landmines in the bush and launch blind ambushes in the forest; human life seemed entirely expendable.

The geopolitical stakes, combined with the mineral and oil manna, distressed me no end. They drew a vast international mafia, big companies and individual adventurers of all shapes, kinds and nationalities, scrambling to gorge on the region. It was feeding time for the vultures on the four-headed beast of DRC, Angola, Cabinda and Congo-Brazzaville. As the sun disappeared from Pointe-Noire, you could see torches burning on the Atlantic coast from north to south. Oil from the Congo and Cabinda never seemed to stop gushing.

As secretary general of Elf Congo, my role involved preparing the company's annual meeting of the board of directors in Brazzaville and inviting the president of Elf Aquitaine, who was currently making substantial cutbacks in the group's oil exploration. This is the man who, in one dramatic fell swoop, had got rid of all of our oil assets in the Doba Basin in Chad, with what I felt was a flimsy excuse: that the Doba-to-Kribi pipeline was too much of a burden for the Group's shareholders.

We therefore prepared this board meeting with utmost prudence, and avoided any ostentatious display of spending. The welcome had to reflect the reputation of my director in Brazzaville, and so I had to organise security for our prestigious visitor from the time the Falcon touched down in Brazzaville through to the drive under military escort to the Elf Congo offices in the centre of town, near the Nabemba Tower.

The escort provided by the administration of Pascal Lissouba in Conga-Brazzaville was more accustomed to the anarchic environment of the civil war than to VIP transport, and they didn't bother with the usual forms of professional discretion when welcoming a company director – there were rockets and grenades scattered on the floor of the escort vehicles, posing a very real risk of explosion. With every sharp turn, I perfected the art of seriously bracing myself.

It was not the most ideal set of circumstances for hosting the president of Elf upon his descent from his Falcon jet. We subtly managed to prevent him from seeing the munitions strewn around the escort cars, but he could scarcely miss the armed men gesticulating from the back of the vehicles that accompanied us, honking and waving their arms, advertising to all who witnessed them that we had the director of a lucrative oil company in our midst.

Every member of the board of directors had accepted the invitation to the meeting, which went smoothly in the duly prepared and inspected room. But just as we were about to have a post-board meeting meal together, suddenly we heard the deafening sound of heavy artillery booming from the neighbouring hills.

The president turned grey and rose from his seat, excusing himself to use the toilet. On his return, looking pale and rather shaken, he informed us he had an urgent and unexpected meeting in Libreville, an hour's plane ride from Brazzaville; clearly he had decided to duck out. He turned on his heel and asked me to take him to the airport immediately, leaving all the guests we had invited to the meal rather baffled about what was going on. He fled without asking me about my family, who were back in Brazzaville, with the cannons getting closer. My mind went back to my encounter with the gorilla; I couldn't help but wonder what would have become of this man had he been in my shoes, given his instinct to flee and to panic. It seemed the company was now run by the type of person I held no respect for, and if that was the case, the future of the Elf Aquitaine Group seemed for me a lost cause. I concluded that it was time for me to part with the company I'd worked for loyally for the last twenty years.

So, I was at a crossroads. I had reached a defining moment in my life, and an irreversible decision had to be made. In the months that followed, a vague feeling quickly came into focus and took on strength, like mounting anger. As the USSR satellite countries were dismantled, the intensity of the space spying conducted during the Strategic Defence Initiative or "Star Wars" period of President Reagan's era dwindled. The last gasps of the Soviet empire in Africa were played out in the civil wars throughout much of the continent, but the threat of a nuclear conflagration between the two great powers had faded.

The result was that NASA suddenly made public thousands of declassified images from space; the peripheral fighting in Africa was of no interest to the major powers, despite the millions of casualties. In the midst of this turbulent period, in 1994, a mining expert contacted me in Brazzaville. He introduced himself as an agent of the CIA and showed me the first radar images of the Congo acquired by the American space shuttle. The aerial images showed very clearly the secret trails that had been cleared in Congo-Brazzaville in order to allow weapons to be carried through virgin forest from Gabon.

The civil war was seriously impacting my work and I decided to turn away from oil and regional politics to resume the mining engineering work I had trained for. These new radar images were extraordinary; they opened my eyes to a new angle in my profession as underground explorer-geologist. Now the answers to what lay underground were coming from another exciting source, and with the innovative technology emerged a new chapter in my professional sphere and a new life. It took me two painful years to bring about this change in direction, and to leave Elf Aquitaine to create my own natural resources exploration company.

More or less as soon as soon as I moved on from Elf, my first radar survey was begun over the Congo Basin. I had had the first glimpse of the results already, during the civil war. This survey was commissioned by a consortium of three companies, Agip, Exxon and Elf, and discussions were underway with the director of Agip Congo based in Pointe-Noire. The initial results were promising, and led to a new oil discovery twenty years later in the targeted area, but that is another story.

6: THE BROODING AURA OF GADDAFI

Two years later, the research centre director at Shell Oil Netherlands called me: he was intrigued by my radar work in the Congo. He asked me to come in and test my new radar tools as part of their "Game Changer" programme, which was devoted to identifying techniques that could change how hydrocarbons are found and reduce the exploration costs. The testing grounds were first Mozambique, then Iran – still under embargo – Iraq, and finally Libya. In Mozambique we were to explore the gas fields in the Karoo, then the Zagros Mountains that run between Iran's western border and north-eastern Iraq, then the Sirte Desert in Libya.

My mission for Shell in the Sirte Desert was to find oil; instead, I discovered water, and lots of it. We uncovered an absolutely enormous leak, several billion cubic metres of water; the scale of it was so enormous, it was initially a mystery where it could be coming from. Then we saw it: this water was emerging from a colossal breach in the huge buried pipeline built by Libyan president Muammar Gaddafi, to move the underground water tapped at a depth of 3,000 metres in the Nubian sands to the major cities in northern Libya.

This great river, known as the Great Man-Made River, was the result of an exorbitant project that harnessed overwhelming engineering resources for more than ten years, and had a budget of several billion dollars. The project, approved and coached by UNESCO, at one point brought glory to Gaddafi. Its objective was to bring drinkable water through a pipeline network over 4,000 kilometres in length to supply the entire coastal population. Despite the prohibitive operating and maintenance costs, the project was a real feather in Gaddafi's cap. The deep water, discovered by oilmen in Nubian sandstone nearly three thousand metres thick, was abundant and pure. This deep water that is replenishable and not salty contradicts the assertions of many hydrogeologists throughout the world, who systematically deny the actual existence of this water. Assertions such as these would have an impact on my work in the future. The several hundred billion cubic metres of it meant that the water requirements of the whole of Libya would be met for many, many years to come. I drank this water in Sirte and in Tripoli, and it was deliciously fresh and good; the fact that it was provided free of charge for the entire population was a fantastic and unprecedented luxury.

There were, however, some challenges to be addressed – this vast source of water wasn't the godsend it may at first have seemed. The water was taken from three sedimentary basins in Tazerbu, Ghadames and Marzuq, and each of these is replenished differently; the water in some of them dates from rainy periods in prehistory, and rainfall has *greatly* decreased. It now only rains once every five or ten years, and very little of that infrequent rain even reaches the ground before it evaporates in the hot air.

Moreover, these huge underground water reserves are not, of course, located conveniently under only Libyan soil: they are transboundary and lie in Algeria, Chad and Egypt, too, and consultation between the countries was not forthcoming. Pumping in Libya for exclusive use by Libya, with UNESCO's blessing, would inevitably have repercussions on the reserves for each neighbouring country, and there would soon be potentially dangerous political consequences related to "ownership" of the water.

Though the pipeline is only a few metres underground, any leaks – caused by the poor quality of the concrete ducts – are immediately absorbed deep down in the desert sands. It is extremely hard to detect these leaks on the surface, where they leave no trace. What could I do with these results? How could we persuade the Libyan authorities to remedy this ecological disaster?

During the winter of 2002, in another unforeseen twist of fate, the president of Congo-Brazzaville, Denis Sassou Nguesso, asked me to accompany him to Tripoli, Libya for negotiations with President Gaddafi concerning an oil contract in the Congo Basin. A few years earlier, the Congo's president had closely monitored the radar work I had done on the basin.

Sassou Nguesso was a strategist who clearly understood the advantages of discovering the oil potential in the Congo Basin. Though it was hemmed in and far from the sea, it was like the oilfields in Chad in the Doba Basin, which today are connected to the Kribi port via an oil pipeline. Discovering oil in this area would shift the geopolitics among the neighbouring countries of Cameroon, Gabon, Central African Republic and DRC, opening up important economic, agricultural and mineral zones all along the Congo River. It would cause a revolution that would bring about, in the long run, a balance with the Kouilou coastal zones, which produce all the oil in Congo.

Our arrival in Libya was welcomed with aplomb – red carpet and cannon blasts greeted us as we descended from our plane. An armada of black Mercedes swept us through the city of Sirte, sirens wailing and horns honking, and Gaddafi awaited us under a big camel-hair tent near the beach, on the sands of the Gulf of Sirte. As the two presidents conversed in the tent, I was left surrounded by bodyguards who stood in front of a huge armoured car right in front of us. President Gaddafi travelled exclusively at night to avoid surveillance – it was apparent that he was constantly afraid of assassination.

Later that afternoon, as I paced up and down the beach, I was joined by Colonel Moftah Missouri, Gaddafi's official interpreter and adviser. He had decided to keep me company and also wanted to set me straight, concluding that I had a Western education corrupted by capitalism, the media and the forces of the colonial empire. Colonel Missouri was an expert, talkative and meticulous. As we walked the beach, he painted an epic fresco of the history of Libya, from Cyrenaica to Tripolitania, and then moved on to the Italian colonisation, as revised and edited by the current regime. He shed a virtual tear or two for

Nazi Germany, then reiterated the Stalinian paean to the phoenix after the coup d'état against King Idris of Libya.

President Sassou Nguesso's delegation was invited to a banquet that evening, given in a sumptuous round building entirely clad with white Carrara marble and embellished with red and green porphyry. Built in the neo-Mussolini style, in the heart of the desert, it was lit by a sparkling chandelier made of several tons of gilded copper and crystal. Some five hundred high-ranked Libyan officers dressed in white, their chests covered in medals, were also guests at the banquet. They sat together in straight rows and were icily silent, in contrast to our noisy and exuberant delegation from the Congo; the six of us had plenty to say. There was also an orchestra, waiting for a signal to start up the music. I was conscious that I was being scrutinised by 500 pairs of curious eyes, all trying to ascertain who the white European was among all the black Congolese.

Then, amid hushed silence, the two presidents took their seats at the banquet, side by side on a high dais facing the assembly. The orchestra played the first notes of the Libyan national anthem, and everyone respectfully stood up. At the end of the meal, as the two presidents left, signalling the end of the evening, a bodyguard blocked my passage and made it clear that I was to follow him, an imperious index finger pressing into my chest. In an adjacent room, President Gaddafi and President Sassou Nguesso were waiting. Gaddafi was sphinx-like and had a furrowed brow. The room was hot and airless – I hoped that the sweat on my face could be explained away by the temperature, but it was incredibly intimidating to find myself face to face with both presidents. As I placed my finger on the red button that started the overhead projector to begin my presentation to the two men, I felt to my horror the barrel of a bodyguard's revolver pressed into my back. I held my breath. The assassination of Commander Massoud in Afghanistan had happened recently, and was still fresh in everyone's mind. The Congolese president intervened with a loud laugh to defuse the atmosphere; the nervous bodyguard backed away, to my relief.

As all this took place, Gaddafi sat silent and brooding, surveying the scene in front of him with a bored detachment. He showed no interest in the meeting, in me, or in hearing what I was about to say; in this unwelcoming atmosphere, it took a lot of fumbling before I finally got the projector working. The first image came on the screen, showing the great leak detected by radar on the artificial river, and the reaction from Gaddafi was instantaneous: suddenly he was shaking off his indifference and feverishly taking notes, to the delight and relief of the Congolese president.

We demonstrated to the Libyan leader that with these powerful new spatial technologies, we could detect underground leaks without ever going on site. My naive assumption had been that by demonstrating the power of our technology, I would get a surveillance contract from the Libyan authorities to monitor the immense underground aqueduct and pinpoint several other leaks. But Gaddafi's reaction was anything but receptive. Despite his curiosity apparently being piqued, it was not in the direction we had hoped; he observed our pictures with cold eyes, and listened to our words intently while silently exuding his disapproval. A few weeks later, we learned that the entire hydraulics team in Libya had disappeared from Sirte.

After a very uncomfortable and agitated night, having stupidly barricaded my bedroom door even though it had no other exit, the next morning I went straight back to Brazzaville with the president, at his suggestion and under his protection. At any rate, it was out of the question that I return to France via Tripoli. I had arrived in Libya without a visa but, above all, Gaddafi's silent disapproval had frightened me to the quick.

After our return to Brazzaville, I began to ponder on something that was germinating in my mind: if I had been capable of identifying an underground water leak in Libya, I should be able to discover aquifers anywhere else. After all, that's what an aquifer essentially *is* – nothing more than one big deep-water leak. But there was no point in looking for water. It had no market value, so what was the point? If I found water, who would pay for my research? Driven by this feverish idea, I devoted a good amount of my free time and my own funds to pursuing this dream. For the next two years, all my weekends and personal finances were devoted to deep underground water.

Water is noble and vital. Sourcing it presented an intriguing scientific challenge, and my intuition told me it was one that was worth facing. I cast my mind back to my experiences in the North Sea: I'd found new solutions when those domes of salt blinded us to what was below. Perhaps a similar technique could provide answers when it came to sourcing water?

In a similar way to the problem with the salt domes, images from satellites were giving us shiny echoes of vast dry and rough surfaces due to the obstacles on the surface. All this information hid other shiny signals, which reflected underground water. But how could we differentiate between all these obstacles on the surface – rocks, villages, camels, cars and so on – and the water underground? We needed to find deep moisture signals in order to find deeper sources of water.

I dreamed about it at night. I reviewed all the mathematical algorithms in existence; NASA scientists had been exploring this possibility since the appearance of the very first satellites, so the technology wasn't new. The strategists were not mistaken in focusing on it – the search for water using satellites is a tenacious dream, and the people who find the solution will access unlimited, unknown horizons. It's still the realm of science fiction – an unknown galaxy that exists within the Earth. But who was I to shoulder such a challenge? It was surely an endeavour to be taken on by a politician or another man of power, not a simple scientist and explorer. But somehow, the prospect of success – the prospect of finding a way of looking beneath the Earth's crust and finding reserves vital to human survival – became overwhelming.

WATER IS A MAGIC MOLECULE

Water, timeless voyager in space, is, after hydrogen, the most elementary and probably the most ancient molecule in the universe, dating from well before the birth of our sun. Water has moved freely through the cosmos since the very beginning of time, it has seen the galaxies in the interstellar void and aggregated into ice at the heart of comets. Some four billion years ago water combined with early terrestrial rock orbiting the sun, contributing to the formation of the Earth and all types of crystals and minerals, later returning in the form of clouds of water vapour produced by erupting volcanoes.

The chemical composition of the water molecule was discovered by French chemist Antoine-Laurent de Lavoisier just before the French Revolution. The French thanked him for this major discovery by sending him to the guillotine at the age of fifty. Lavoisier had seen workers producing a combustible gas with steam from a boiler. They did this by simply sending the steam through a red-hot gun barrel that had been heated over burning coal; the gas that came out burned with a bright-blue flame. The strange phenomenon he witnessed intrigued him. He later named this gas hydrogen, as it was generated by water.

Lavoisier's genius was borne out: he weighed a new gun barrel before the heating operation and weighed it again after the experiment. He found that, after having given up its hydrogen, the gun barrel was heavier and was covered with rust like an old nail.

In fact, this experiment quickly made Lavoisier understand that the weight difference came from the transformation of the iron in the gun barrel into rust or iron oxide, and that the oxidation was caused by another molecule generated by the water vapour that oxidised the barrel. Thus, water contained another gas, which he named oxygen. He had just identified the formula of water: H_2O.

This was the birth of modern chemistry and led to the end of alchemy. Physicists and shamans were in agreement – both would describe the water molecule as an extraordinary, versatile and paradoxical divinity. Water in its solid state does not become denser upon cooling, like most other constituents of nature. It becomes lighter. Ice does not flow; it floats on water.

Ice under pressure does not become more solid either. Unlike anything you'd expect to see, it goes back to a liquid state: ice melts between the jaws of a press, just like it melts at the base of glaciers. This enables it to flow, which is quite fortunate. Otherwise, huge blocks of ice would remain eternally blocked at the two poles, totally skewing the circulation of ocean masses and of course the climate.

Imagine Darwin's surprise during his voyage on the *Beagle* between 1831 and 1836. Throughout his five years on board ship, sailing the world's seas, he measured the temperature of the deep water in the oceans. Wherever he took his measurement, he found a constant temperature of 4° Celsius! He discovered what we have confirmed today from physics and thermodynamics: that under high pressure, water is densest at this very specific temperature absolutely consistently. Who benefits more from this property of water than the trout? A single swipe of its tail is much more effective in a river where the snow is melting than it is in a warm stream in the summer. Dense water at a temperature of 4° Celsius offers greater lift for the trout's movements and enables it to dodge obstacles more quickly than it could in warm water. The unified surface of a water table is governed by a force called surface tension, created by van der Waals intermolecular forces, which have the effect of skin – the capillary effect. This forms a sort of carpet and is what enables light insects to walk on water.

Cold water dissolves more oxygen and carbon dioxide than does warm water; this basic property explains the origin of the life that appeared in the sea. Today, the oceans under the polar ice teem with life, whereas warmer waters, though conducive to biodiversity, do not hold such a prolific array. Cold water, with its carbon dioxide and acidity, carved the limestone hills into a sugarloaf shape in Halong Bay in Vietnam or the Dolomite Mountains in Italy. During the Ice Ages, acid-rich and carbon-dioxide-rich cold water also carved out all the caves in Europe and Siberia where prehistoric people took shelter.

As Lavoisier once said, on Earth nothing is lost, nothing is created, everything is transformed. That is exactly what happens in the water cycle.

Potable water, whether from a river, a lake or a well, comes primarily from the evaporation of seawater. All this water vapour in contact with the continents condenses and precipitates in rainfall when it reaches land. Part of it evaporates (40 to 80 per cent, depending on where), part becomes run-off that feeds rivers, and the smallest part (1 per cent to 30 per cent) percolates in porous ground and runs into the groundwater table. In the end, at different points in time, all water returns to the sea in liquid form, and the cycle begins again. The water table is by definition underground – the water is captured in rock where it is stored, refilling and circulating freely. Such a formation is known as an aquifer (literally, an underground layer of water-bearing permeable rock).

All the water in these aquifers comes from rainfall, which at one time or another has managed to percolate down into the aquifer and accumulate, to fill and recharge the aquifer either vertically and directly, or through longer, more complex pathways underground, along fracture lines. The simplest underground bodies of water are created by river sand, which creates alluvial aquifers – these are directly recharged by the river and by rain. Generally speaking, the biggest aquifers are in (sandstone) geological formations, which can be very wide, thick and close to the surface like riverbeds. They can be under sand dunes, or deeper, like in ancient deltas. Over several million years of the sedimentation process, these deltas end up deeply buried.

The Nubian sandstone aquifer system is the world's largest known fossil water system. Nubian sandstone is more than 60 million years old and holds underground water beneath four countries in north-eastern Africa: Sudan, Chad, Egypt and Libya (including that exploited by Gaddafi's Great Man-Made River). As revealed by oil exploration, this water spreads across thousands of square kilometres at depths of nearly 3,000 metres. The water is of excellent quality but unfortunately it is not easily renewable, for most of it comes from ancient rain that fell when the climate in North Africa was wetter. That ancient wet climate was very conducive to direct vertical infiltration of the rainwater, leading to the formation of this giant aquifer over several centuries. In the case of these aquifers, we mustn't think of them as big underground lakes, but rather as vast sponges that hold water in their holes. The easiest way to understand this type of aquifer is to imagine a sponge sitting in a big soup plate with water at the base of the plate.

If the space between the grains of sand is too small, the water will be retained by capillary effect and won't be able to flow. The grains of sand have to be sufficiently coarse so that they can store the water and enable it to flow freely. Different wells that have been drilled into this system are more productive in zones with coarse sand, and less productive in zones with finer sand.

Surface aquifers are known as unconfined aquifers: they recharge directly in contact with a river or through rainfall, and flow freely over impermeable land surface that does not let the water infiltrate any deeper. Deeper aquifers are most often confined, for they are trapped between layers of impermeable material above and below the aquifer. They do not recharge directly and their refill mechanism is often complex. Again, a sponge between two soup plates gives a fairly good idea.

There are several differences with oil. Oil floats on top of water. It tends to migrate towards the top of structures in the form of ridges (hogbacks) known as anticlines, rising up to the surface of the land, whereas water tends to migrate downwards and stagnate in the depths of hollow structures (called synclines), which makes it hard to find this deep water.

Other aquifers are known to speleologists, who explore underground galleries in limestone formations eroded by rainfall. These underground holes, called karsts, store water and let it travel as best it can with all the setbacks on its path. It is hard to drill this type of aquifer, and not recommended: they feed rivers, which could then dry up. There is one exception, however, and that is when these resurgences flow into the sea, which does happen in several places on Earth. In these cases, the fresh water can be captured before it dilutes into the salt water. Such work is painstaking and detailed, a fine art that requires science and intuition.

There is one more type of aquifer that is little known and yet has good potential. It arises from a recharge of crystalline basements in mountainous areas, with rainwater penetrating deeply through the dense fracture system. Water that has been accumulated over millions of years can occasionally create giant aquifers that have not yet been identified by traditional research and are out of the scope of traditional hydrogeological drilling.

Such an aquifer, detected by an American exploration firm

and drilled on the island of Trinidad, is currently delivering good flow rates and providing a fresh-water resource. This type of deep aquifer can also flow into the sea; one has been detected off the coast of Ghana, flowing at a depth of 400 metres.

To sum up this infinite complexity, if we could take a trip underground we would see a series of aquifers of every size, in different types of geological formations, interrupted by fractures in the rocks at depths that range from riverbeds at the surface right down to deep synclines equivalent to large underground valleys which accumulate huge water quantities in sedimentary basins at several thousand metres. This elusive buried water is tracked down in steps, by geographic segments, by geological and structural styles inherent to each region of the globe and its pluviometry. A dry country offers less chance for aquifer recharge than a wet country. The situation is further complicated by recent climate changes on a planetary scale over the past sixty years. This global factor is worsening, and makes the search for deep water all the more crucial today.

Buried water travels over huge distances and on a long, slow scale of time. As a result, research into buried water must often be conducted across several countries. The water in a well in the central Sahara in Algeria comes from rain that fell on the Atlas mountains in Morocco more than a thousand years ago, during the reign of Charlemagne. At that time, the area was green and lush, covered with cedar trees that enabled the water to slowly filter underground. Now, the water level of these wells is regressing in the thick sandstone deposits of the Tertiary age of the Sahara, and the cedar trees in the Moroccan Atlas were cut down long ago . . . Which raises the question of the impact of worldwide deforestation on the recharge process of underground aquifers.

7: FEET IN THE MUD, HEAD IN THE STARS

"All grown-ups were once children . . . but only a
few of them remember it." Antoine de Saint-Exupéry,
preface to *The Little Prince*, addressed to his best
friend Léon Werth "when he was a little boy".

Astronomers had addressed the problem of atmospheric wave perturbations by installing the Hubble telescope in space so that they could better observe and understand the clusters of faraway galaxies, and it struck me that this was the direction to take. I was facing water many metres deep, and I needed to find a way of getting through the surface obstacles and the first few metres below the surface in order to be able to see only the signals that related to the water deep beneath, our underground galaxies.

It struck me that there must be a solution, and the quest to find it occupied my mind for more than two years. I tested algorithm after algorithm for mathematical processing of the signal, reminding myself of what I had once achieved as a young engineer at Elf Aquitaine in the Chambourcy research centre, but each time it was in vain. Day after day I eagerly worked on countless calculations to make progress in my vision of the underground, groping in the dark, using trial and error. Weeks could go by with no notable progress, for the surface obstacles only disappeared with the signal of moisture, precisely the opposite of what I was trying to obtain.

I was on the verge of giving up when, one night, a solution crept into my subconscious as I dreamed – it seemed miraculous but, really, the answer had clearly been brewing away under the surface for some time, the various attempts that had ended in failure eventually adding up to the answer. As unlikely as it sounds, a dream brought me a solution and I scribbled it into a notebook, fearful of not remembering it the next morning.

As soon as I woke up, I loaded a final program for image processing that I had just obtained and, after two hours of calculations, an image of the underground I had never seen before appeared on the screen. On this mapped image, all the surface obstacles had been filtered, like with the Hubble telescope. My heart pounded in my chest; here in front of me was an unknown world, with its black holes, its star clusters, its brilliant streaks . . . all clues I could use to interpret what was there beneath our feet. A new underground galaxy opened before my very eyes, in a "sky" that had previously been obstinately dark. This was the magical birth of the WATEX™ process. It wasn't long before my discovery led me to see gigantic reserves of water trapped in rock, in some of the most unexpected places on Earth.

Two months later, in February 2004, I received a call from someone at Spot Image in Toulouse, newly appointed to UNOSAT in Geneva. UNOSAT – the Operational Satellite Applications Programme – is responsible for studying satellite images for humanitarian causes; my contact wanted to talk to me in relation to the High Commission on Refugees in Geneva. Two hundred and fifty thousand refugees fleeing Darfur had recently come to Chad along a 650-kilometre front on the border of Sudan. There were thousands of children, and countless refugees were sick and wounded. Many of them were dying of thirst.

Was there a possibility that we could use our new process to find, quickly, drinkable water for all the refugees fleeing Sudan? It was a heavy responsibility, and it hit hard; these were many thousands of lives, and we were being asked to help save them. The prospect of failure was abhorrent, the prospect of success wonderful. It seemed a logical conclusion to all the trial and error, and other difficulties we'd experienced over the years; I didn't have an immediate response for them, and couldn't reply with enthusiastic positivity, but my intuition told me we had to try, and the water battle began. We couldn't hesitate, couldn't evade the question: each passing day meant more victims.

A day later I was in Geneva developing an emergency action plan with my partners at UNOSAT. I had plunged myself straight into an unknown scenario; it was a race against time, a challenge to prove to international institutions that we had a solution when it came to the mass deaths occurring on the far side of eastern Chad. Somehow it was virtual, still didn't feel tangible; the crisis was too horrifying for the reality to hit home, the cost to life too great.

We had four months to locate the main water resources in an arid country that covers 80,000 square kilometres, roughly the size of Portugal. We immediately set to work examining the entire country, layer after layer, peeling the surface of the Earth like an onion, looking into every nook and recess for water resources.

The NGOs scarcely knew where to begin. They were drilling blindly and urgently. Their success rate reached only 25 to 30 per cent, which equated to approximately one productive well for every two or three dry wells ... So much time and money were wasted, and the number of victims continued to rise, including the drillers. It was vital that our results helped the NGOs focus on drilling wells that had been clearly targeted and accurately identified within ten metres. That was the goal we wanted to achieve, even though we well knew it would not be simple; the desert made everything more complicated, not to mention that it was a war zone, and the trails were studded with landmines. There was no potable water, no food, and no security in the camps, which frequently fell prey to attacks from horsemen and bandits who took advantage of the situation to pillage the region.

Every night, inside the high walls of the Ursuline convent in Provence where I still have my home, my office and my employees, billions of pixels were downloaded from different space agencies. We reassembled optic images and radar scans in the form of mosaics of images, block by block, with geographic coordinates that have an accuracy of within six metres on the ground.

After three months of intensive work on our computer screens to process and read images from NASA, and from Canadian,

Japanese and European satellites, we began to see the presence of some still-unknown aquifers. I was going to have to confirm them by drilling and monitoring the operations in the field in order to calibrate and confirm the efficiency of my new tools. This was almost twenty years ago and we were still in the experimental phase, at the beginning of a great scientific and human adventure. My oilman's intuition was telling me that we were on the right track; I was determined that we would succeed.

Three months later, I jumped on a plane and flew straight from Paris to N'Djamena in Chad with just a backpack, hammer, GPS and my underground navigation system loaded with the new WATEX™ image. Everything we were looking for was deep underground and could not be seen in the surrounding countryside. I had to couple the digital image of the buried aquifers loaded on my laptop to a GPS, which would then tell me precisely where I was, like a pilot navigating in fog who has accurately located, by radar, the runway on which he has to land.

In the meantime, we were receiving alarming news about multitudes of refugees who were dying of thirst daily along the Sudanese border. The UN Refugee Agency (UNHCR) recovered countless people in desperate physical condition, many of whom died within hours despite intensive rehydration efforts. These awful snippets of news reminded me of my survival courses in Holland, where we learned the hard way how to stay calm if our helicopter was to crash into the sea. We were taught that pilots shot down in the North Sea inevitably died after ten minutes of having been immersed in its 4°C water, even if they were alive when fished out.

Techniques of torture involving thirst have existed from time immemorial, and bring about atrocious suffering; but what is really going on in the body? Water is the primary solvent that regulates our body. Every single bodily function depends on water; they also depend on other mechanisms such as blood flow, nutrients reaching our cells, evacuation of dead cells, circulation of nerve pulses and hormones throughout the body – and of course, all of this is controlled by the workings of the brain. Just for normal functioning of our metabolism in a temperate climate, we need to drink two to three litres of water every day; in a desert climate, we need five to six litres per day just to maintain our body's status quo.

With the onset of dehydration, the blood begins to concentrate and the flow of it slows down as it thickens; there is a painful burning sensation in veins and arteries. Then, gradually, the tongue sticks to the roof of the mouth and swells. The eyes become dry, the vision rapidly deteriorates, and the ears start humming. A person can no longer talk and soon they can no longer see. The heart starts to beat rapidly, racing in the chest, because it is working overtime to try to push the thickened blood through an exhausted body and shrinking arteries. Thirst then begins to torture you, and headaches set in. Blindness occurs, as do painful muscle cramps. This is the point of no return, with the start of a fatal chain of events: the kidneys fail, which causes uremic encephalopathy, and cardiac arrhythmia sets in. Blood pressure climbs and blood vessels burst in the brain. It is a slow, agonising death under the relentless heat of the sun, and it ends in silent convulsive spasms. The urgency of the situation was at the forefront of my mind, and yet the challenge seemed impossible. I felt a terrible sense of responsibility and a glimmer of hopeful optimism that perhaps I could make a difference.

Above: Clearing hacked out of the forest with chainsaws to allow a helicopter to drop seismic teams and their equipment.

Above: Water vines in Gabon primeval forest, in the Lambaréné area: the healthiest way to quench your thirst.

Below: The author waiting for the helicopter on a helipad in the Gabon forest with his Pygmy friends: machete-wielders, compass-readers and dynamiters.

Left: Tough crossing through a deep swamp with the help of the Pygmies, trying to protect my bag and notes.

Above: Barricades in front of the Soviet parliament on 22 August 1991.

Left: With my interpreter Iouri Kouchnariov in front of the Kazakh Council of Ministers on the very day of the coup d'état, 19 August 1991.

Right: Radar image of the great leak covering several thousand hectares of land in the Sirte Desert, Libya.

First WATEX™ image produced in Darfur by RTI, between Chad and Sudan, compared to a Landsat optic image. The unknown world of underground water is revealed by yellow and blue trends, with its black holes and star clusters.

Left: Animals drop dead of thirst all along the Chad border: they are all that remains of the worldly goods of the Darfur refugees.

Darfur refugee children in a sandstorm on the Chad/Sudan border, awaiting help, in July 2004.

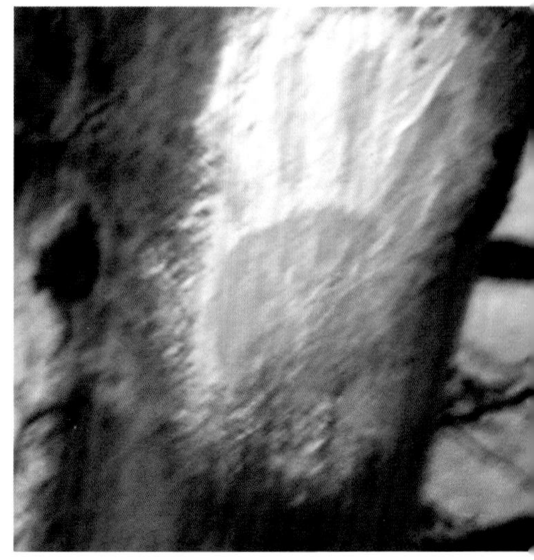

The yellow imprint of my hand on a tree to indicate a drilling point in eastern Chad, July 2004. A sign of hope.

Landmines in the Lobito-Catumbela area of Angola.

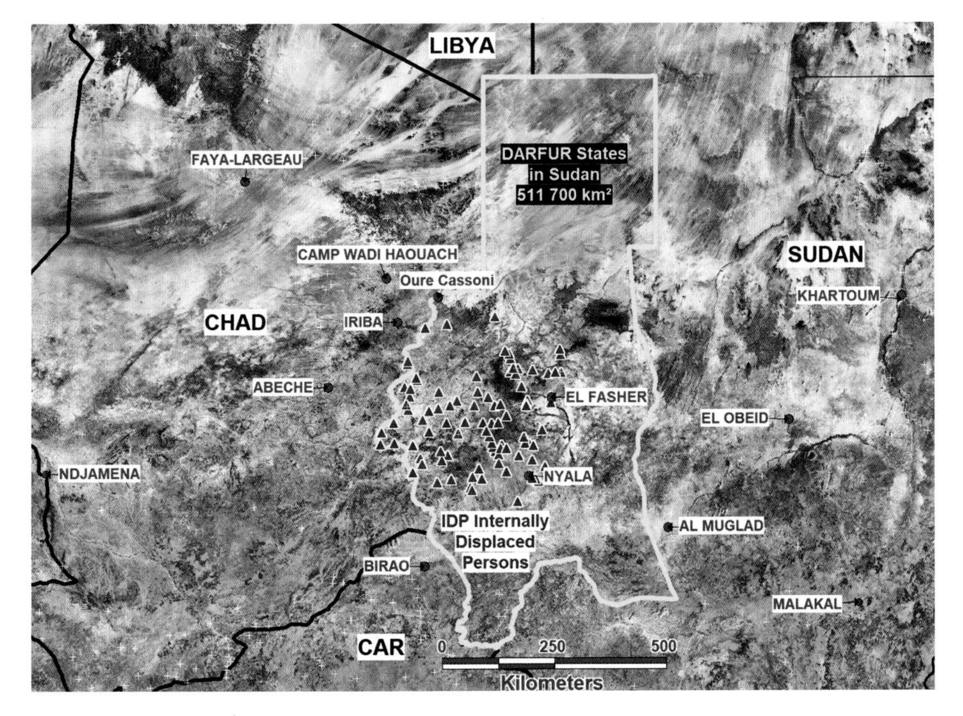

Satellite image showing the Darfur area, the Sudanese refugee camps in Chad (Iriba, Abéché, Oure Cassoni) and the camps for displaced people in Sudan (blue triangles).

Young women at the well in West Darfur, July 2004.

Pygmy prospectors pouring sandy water into a sluice box to wash and extract gold.

Above: The Ennedi plateau between Iriba and the Fada oasis, taken from my plane window.

Right: Prospecting at the base of the high Ennedi cliffs, under the protection of the MINURCAT. (Photo Alain Gachet)

Negotiating in front of the APC tank. L to r: Alain Gachet, Haroun Tagabo, Mongolian Major Ba'Atar.

Above: Strategy in the tent of the MINURCAT Mongolian batallion at the Haouach camp. L to r: prefect Issaka Hassan Jogoï, Alain Gachet, Mongolian Major Ba'Atar and Tunisian captain and UN observer Mamdouh.

Left: Mongolian soldiers from the MINURCAT preparing the barbecue.

The author working on geological prospection escorted by Gansukh, the Mongolian aide-de-camp, who was also the unarmed UN observer along the Ouadi Haouach.

A young Goran girl from the Toubou people, riding her camel to pull the rope and lift water from the well to water the herds.

Turkana orphan digging into the dried-up riverbed, looking for water for her two goats.

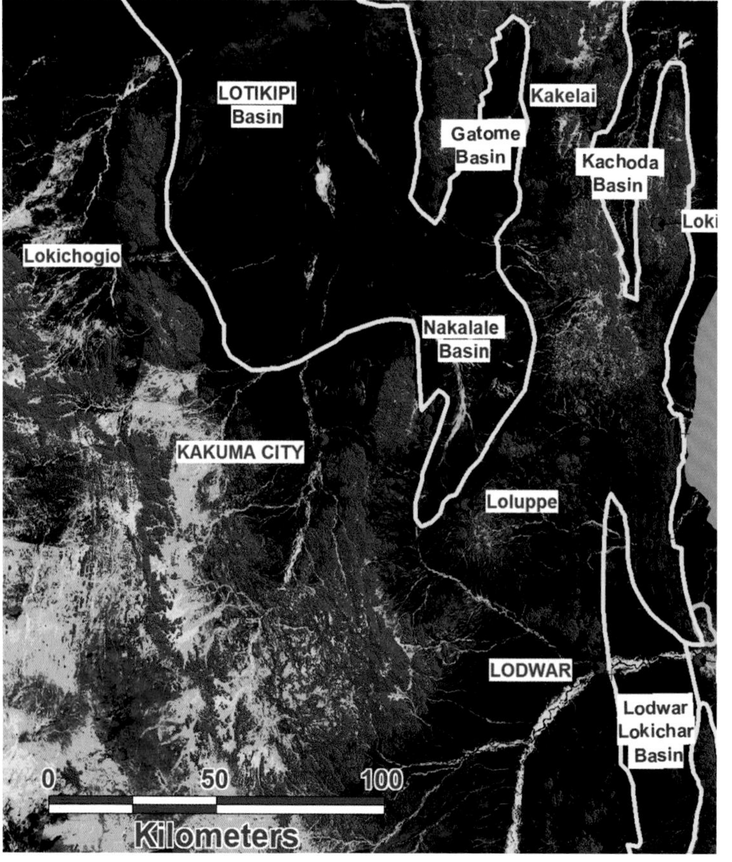

Left: WATEX™ image of Turkana revealing five giant aquifers (yellow polygons) in the black holes of the underground galaxy: Lotikipi, Gatome, Nakalale, Kachoda and the Lodwar Lokichar Basins. © RTI

8: BETWEEN PRIDE AND LIGHT IN DARFUR

Landing in Iriba, a remote settlement in eastern Chad, in July 2004, was a moment of reckoning. The frail twin-engine plane chartered by the United Nations dropped me in what seemed to be the middle of nowhere, on a short laterite runway 850 kilometres north-east of N'Djamena, Chad's capital. The dusty little village of Iriba is the last village before the border with Darfur in Sudan, some sixty kilometres away.

It is a humble little settlement with a few dozen adobe huts with corrugated metal roofing huddled around the house of the sultan, a vast concrete building painted white. We could see that there had been no food or water for a very long time; at what is still referred to as the "market", there were a few cloves of garlic and some corrugated tins of food that had long since expired. The people themselves were in a terrible state of physical decrepitude; the famine had clearly been rife in the area for many months.

The American embassy in N'Djamena had given me a few MRE combat rations (Meals Ready to Eat) that I later shared with my newfound companions, and within a few hours my status had changed: the refugees were clear that I was on their side, the fact that I had a backpack and a pump to purify whatever water I hoped to find in the ditches or in the wadi perhaps a promising sign, rather than something to fear.

The UNHCR base camp in Iriba was set up in a large hut surrounded by an adobe wall in the centre of the village. We shared a big room that teemed with insects and scorpions, centipedes, roaches, locusts and spiders including tank-carriers, a type of large hairy spider that, rather revoltingly, carries baby scorpions around on its back. Despite fierce and energetic sweeping, our room remained the Tower of Babel for every chitinous and crawling insect on Earth. The camp guard did however point out how lucky we were not to have snakes, which was quite reassuring, though there was always the risk that rainwater would later chase them out of their snake holes, so, the sense of relief didn't last long.

There was an absence of insecticides and so our tolerance became relatively high; however, I had never dreamed that these swarms of insects would noisily intrude on a permanent and perpetual basis. As we lay awake at night, the noisy crunches were impossible to ignore; the roaches had loud footsteps, as though a person wearing clogs had suddenly entered the room. Try as we might to ignore it, it was a cacophony of sound that never seemed to end. It was all around us, too, on the floor, on the walls, on the ceiling above our heads . . . the invasion became a kind of haunting. We did our best to laugh about it – what else could one do? – but getting up at night and setting a foot on the ground only to hear the crunch of countless insects soon drained the situation of what little humour it may at one time have possessed. Small holes in our mosquito nets let aggressive insects into our sleeping bags, and we were covered in angry bites. These tiny beasts weren't attracted by our crumbs – we had next to nothing to eat ourselves – and the hungry vermin had no other choice but to eat each other. Before the first big sweep in the morning, we couldn't resist the fun of counting the number of half-devoured cadavers scattered on the floor, which gave an insight into the insects' nocturnal fights.

After a restless night of interrupted sleep, I realised that the old lurching Land Rover rented from the sultan of Iriba had no spare tyre. I enquired about it to the driver, a distant relative in the sultan's vast family, who replied with total confidence that I wouldn't need a spare tyre because "tyres never go flat around here!" His assertion seemed a bit nonchalant for my tastes, and I requested a spare before leaving. It soon dawned on me that I would have to go and greet the sultan; indeed, that I was expected to bow down before him. After the ceremony, which was required by local etiquette, I climbed into the back of the car and found the precious spare on the back seat. That was a good start, but there was no jack or lug wrench to go with it . . .

We wasted a lot of time that first morning with one thing or another and, when we finally got underway towards the east, the blazing sun beat down on land that had been ravaged by drought. The driver took us to the village of Tiné, in the immense desert and dunes of the Eastern Ouaddaï region. As we arrived near the Sudanese border, we saw groups of refugees walking along the track, horrifically emaciated and nearly blinded by the sands. They walked slowly in a column, heads down; they were silently dying of thirst. What made the horror of what we saw so distressing was our powerlessness. We could help, perhaps, in the long term, in a bigger way; but I couldn't look at them, such was my shame, my horror. Sandy winds lacerated the gaunt black basaltic landscape traversed by white sand dunes and rocky outcroppings. It offered some relief to witness UNHCR trucks recovering the straggling refugees one after the other, to take them to hastily assembled transit camps.

To say that July in this part of the world was hot would be an understatement; temperatures easily reached 50°C/122°F, in the shade. The road running from Iriba to Sudan was studded with hundreds of corpses of dead donkeys, sheep, goats and camels, abandoned along the track. They were lined up in a peculiar imitation of a polite single-file queue: it was a physical manifestation of pain, horror and poverty, snaking relentlessly through the dunes. The sand soon covered the corpses like a discreet shroud.

The dead animals along the Sudanese border reflected the lost fortune of those lucky enough to survive. The wealth of thousands of refugee families lay exposed under the relentless sun, in an advanced state of decomposition. It was a clear picture of the total ruin of the Darfur refugees fleeing Sudan under cover of sand-laden winds; there were so many corpses, even the hyenas no longer bothered.

That very same morning we had been laughing as we counted the number of insects in our room; here, there was no humour, and I couldn't count all the cadavers in the plain. Behind this apocalyptic vision were many more human victims; the overwhelming vision of death in front of me was unbearable. I clenched my jaw and told myself that the best way to help these refugees was not to take them in my arms, or cry over their fate, but to find water, and find it quickly. There was nothing I could do to help with the kind of immediacy I longed to, but perhaps, long-term, I could change this dreadful outlook. I gave instructions to my driver without looking at the landscape around me.

I had never before been to this area, but I had analysed it by satellite, I knew every nook and cranny – of the underground area at least – and I managed to locate in real time, by GPS, our position on the WATEX™ map on my navigation screen as we drove through this scorched, desolate place. I had to cover the car windows with towels to get rid of the glare and be able to see the images on the screen as we advanced towards Sudan. It was also a way to avoid looking at the ghosts that lay around me, and which would later come back to haunt my nights.

I gave very simple directions to my driver: "right, left, straight, stop"; these were the only instructions he needed to arrive at the buried targets that were invisible on the surface. With my maps and my GPS, we finally reached the village of Tiné, a settlement on the Sudanese border. On arrival, it was clear that it had just been bombed by Sudanese planes. Groups of haggard and disoriented refugees wandered like lost souls on the road to Iriba, looking for who knows what salvation. One by one, like others before them, they were picked up by UN trucks to be taken to Bahai and Cariari camps, further north. Had they attempted to walk to Iriba they would have died of hunger along the way, like their livestock, which we could see all around us.

Those who had lost their children took care of the orphans. They rinsed their eyes and gave them sips of water from their meagre rations. They moved onward without a complaint, their villages burnt to the ground, the blinding sandy winds offering them some protection from the Arab horsemen who'd set fire to their huts. Their enemy had also tossed the cadavers of their families and animals into their wells, poisoning them. The destruction was methodical and relentless in its cruelty. The population was terrified; the fear in their eyes made my blood run cold.

Moving forward with the wave of refugees towards the north, we zigzagged along the border for the rest of the day. We were heading for the Bahai and the Oure Cassoni camps, crossing battlefields and swerving left and right to avoid the spent munitions and unexploded shells sticking out of the sand. Though we didn't know where the camp was located, all we had to do to find it was follow the black plastic bags caught in the branches of the acacia trees and continue upwind. NGOs distribute food to thousands of refugees every day, in these thin, light black plastic bags that then blow into the trees for dozens of kilometres all around. I saw a compact mass of the bags in the stomach of a rotting camel by the roadside.

We stopped at the water targets I had identified by satellite, and, armed with a can of yellow paint and a brush, I marked the appropriate rocks with the yellow print of my hand.

In the painful aftermath of what I'd just witnessed, I began to contemplate how absurd my situation was. What was I doing here, a petroleum engineer with a family back in France, in the midst of death, and stench, and suffering? Staying seemed absurd, but so did the concept of returning to the creature comforts I'd come to take for granted. It was ludicrous to be here; it was ludicrous to leave.

The importance of being in this hostile environment, where the yellow signs of survival I was painting on the rocks at the right places would lead to finding potable water for these people, was palpable. I couldn't guess at the time that I would find water elsewhere on Earth; my future discoveries were an irrelevance. I sensed very strongly that I was in the right place at the right time, and that here I could try, at least, to help these thousands of people to survive, to get beyond the terrible disaster they had found themselves in.

We had no time to be afraid. Our work had to be done quickly and efficiently, to minimise exposure to accidents and to hostage-taking. Extreme tragedy has a tendency to take you by surprise – we were contemplating tangible accidents, like injuries, mutilations and death, but in fact, fear was all around us, and manifested itself in different ways, which were often subtle, ordinary even; when fear took over, there was no getting rid of it. Fear could be debilitating and paralysing – we did our best to combat it. An odd kind of "courage" emerges in situations like this – it's not an overt bravery, borne out of a lack of consideration of one's own safety, or selflessness;

instead, it's a kind of resignation to the universe. Not an admission of defeat, or a blind acceptance that something awful will inevitably happen; rather a sort of trust that the only answer is action, the only solution is to keep going when you can do good.

I would be facing this situation later in Angola, near Catumbela in the province of Benguela. Landmines had been planted everywhere, and we could only travel along the very narrow corridors that had already been cleared of mines. I continued to worry more about the idea of delivering water to these war-torn populations than about the danger of landmines and kidnapping inherent in this situation.

Late that afternoon we arrived at the Oure Cassoni camp near the Cariari dam at the border with Sudan. The noxious, murky pond, almost dried out and stagnating at the bottom of this artificial reservoir, supplied a makeshift camp for 25,000 people. There would never be enough water in the upcoming weeks to supply the refugees streaming in, and tank trucks containing drinking water were already on their way. I met up with the UNHCR members on the highlands, where thousands of tents had been erected. They were there to handle emergencies, record the arriving refugees and allocate space in a tent to each family. The UNHCR handed each person a badge that gave them access to food and medical supplies.

Violent gusts of evening wind had kicked in, a sign of an impending sandstorm or dust storm, perhaps even indicative of rain. The air was thick with a dark, fine, choking dust. I was discussing the camp's water needs with Geoffrey Wordley, one of the UNHCR leaders, when a deafening scream came from a woman whose tent had been ripped in two by the violent winds. We both rushed to help, grabbing at the tent pegs and attempting to push them back into the ground, but it was too compact, too hard. Fortunately, I had my geologist's hammer to hand, and we used it to drive in the ripped-out pegs. We set the tent straight, our eyes gritty with sand.

The woman continued her lamentations, raising her arms to the Lord. We checked her meagre belongings and her two children inside the tent, and all seemed well. She then addressed us in English, chanting a long litany with great conviction: her husband was rich, very rich, he had four other wives and would soon come to rescue her from this disaster. But where was her husband? Geoffrey told me he had been killed near the border with all his other wives, and this fifth wife was in the process of losing her mind, her two children curled up like little animals inside the tent.

As night fell we left for the Bahai camp, a few kilometres further south, on the Sudanese border. We hoped to find the UN protection teams and a good place to spend the night.

At Bahai, a makeshift school had been dismantled; all that was left was a corrugated metal roof laid on top of collapsed adobe walls. There we found the NGO members who worked at the Oure Cassoni camp. The luckiest ones had found space under the big UNHCR tent. The fifty young workers who volunteered for a variety of aid organisations crowded together under the sheet-metal roof, where muddy drops of rain were falling heavily from dark clouds streaked with lightning.

Spiders, centipedes and scorpions, flushed out of their flooded holes and nests by the pouring rain, invaded our sleeping quarters and drove some of us outside. After a few minutes, the rain became lighter and cleaner. I took advantage of the fresher water to fill up the water bottles in the room from the run-off dripping from the metal roof. It was the only way to refill our water supplies for the uncertain days ahead. In my pack, I had two Havana cigars (I always keep these on hand for particularly stressful times), and I shared them with Geoffrey, my comrade-in-arms, in our setting of total destitution.

The rain finally stopped and the clouds cleared. The night was clear, quiet and cool. We did not want to share the sleeping space with the swarms of insects and instead decided to create a camp inside the school. We set up our cot and mosquito net, then opened a few cans of our rations. Both of us felt uneasy – we were less than 100 metres from the border, with no protection from Sudanese attackers; we were also near the dry bed of the Wadi Howar, the starting point for Arab horsemen launching raids in Chad. Here they took hostages and kidnapped the young women in the refugee camps to sell as slaves to rich Arab landowners in Sudan and in the Arabian Peninsula. Despite my high adrenaline level, I finally fell asleep, hopeful that the holes in my mosquito net would not allow too vicious an insect onslaught. Let happen what may – we had no other options in any case, and tomorrow would be a long day. If indeed the threat of attack from nearby Arab horsemen would allow us to live another day ... who knew? We were too tired to even worry about that.

The next day, I continued to paint white crosses and affix yellow hands to rocks to identify the most pertinent drilling sites. As we advanced, we saw several emaciated donkeys on the track, sagging under the weight of bricks they were carrying in wooden baskets. Their master struck them, landing harsh blows on their backs and sides, yet they continued faithfully performing all the work imposed on them. I asked my driver who fed them in exchange for the hard work they did, given the shortage of food in the region. My question surprised him and he answered with a defensive edge to his voice: "No one feeds them, we let them go into the bush and they have to make do for themselves!"

I retorted, "But if they work for you, you should feed them. Otherwise, they're going to die of thirst and starvation."

His judgment brooked no interference: "If they die, it's their fault. Too bad for them. So what? There's plenty of donkeys around here, we'll just take another one ..."

In some strange association of ideas, he then continued on the subject of women. He seemed to have well-established opinions about women that were very similar to his comments on the donkeys. His first wife had come from the capital, N'Djamena, and expressed, it appeared, a lot of wishful thinking about independence, due to her dissolute moral environment from being in contact with the big city. He had tried to put a stop to her thinking, by "striking her hard with a big stick", but that didn't work either. In the end he had felt he had to get rid of her – she was "untamable". He then chose his second wife from a Zaghawa tribe in Iriba, where submission is the rule, under the authority of the sultan of Dar Zaghawa, known locally as Bakhit Hagar. He still had to strike her regularly, to remind her who was the master, but that seemed to work; he seemed to be almost fond of this one, or at least had no plans to "return" her. He wanted also to take on a new, second wife but was lacking the financial means to make that a reality. Regardless, when the time came, she would have to be from Iriba or Abéché; certainly not from N'Djamena, given his disappointing previous experience.

As we made our way along the track, I continued to mark drilling sites near the Sudanese border. I was down in a dried-out wadi when a big black car pulled up. Out of it stepped a frail white man with a bald head and a big smile; he approached me and extended his hand, but mine were sticky and covered in yellow paint, and I held them up apologetically. He was American. He asked me what I was doing there in a ditch, with my pot of paint and my GPS, why I was in this danger zone with no protection. I explained I was looking for water for the refugees, and marking the position of future wells. He seemed intrigued by my evasive response but I had no time to talk. He offered to meet me that same evening at the Iriba camp, which seemed a better idea to me. At least I could just about wash my hands, give him a yellow-tainted business card and get his; it was a slightly mysterious exchange and I was intrigued to find out more about him.

In the late afternoon, as we were returning to the camp in Iriba, my driver – the one with extensive knowledge of donkeys and women – managed to get lost and go down a track pockmarked with landmines. Mine-clearing teams had left very explicit signs: yellow tape and white skull-and-crossbones on red background. It was 6 p.m. Night was falling and armed camel-drivers could be seen lurking in the desert. I suddenly felt extremely vulnerable, and clearly sensed the hostile world around me. No water, no food, and the risk of losing my life or, even worse, surviving with lost limbs in this minefield; and my driver wasn't the most ideal companion. If I was going to lose my life next to anyone, it seemed a shame it had to be him.

We had to leave this track and take a roundabout route along the Abousnet wadi, a dried-up riverbed with jagged banks with spikes of sharp basalt. We also had to keep as much distance as possible between us and the border with Sudan. My aim was to find the Tiné road in Iriba, in total darkness, without rolling the car over on the rough banks of the thalweg or being devoured by hyenas, which proliferated in that area. Around midnight, with a bit of moonlight and unhoped-for luck, and the precious assistance of our GPS, we emerged from the jagged surroundings. When we arrived in the Iriba camp, around 2 a.m., the UNHCR was panicking; they understood we were in trouble but the thought of calling them for help was simply ridiculous as, despite their best efforts, they would never find us in this chaotic mass of rock and sand in the dark.

Slipping into my sleeping bag that night under the mosquito net over my cot, I felt weary and aged, exhausted by this humanity at death's door.

The next morning after a relatively sleepless night spent battling insects, just when we were allocating combat rations for breakfast, I saw the American visitor I had met the previous evening at the wadi. He was pleased to find me safe and sound and listened wryly to my adventure on the Sudanese border that had mobilised the powerless UNHCR rescue crew the night before.

I explained to him how I worked, and that I was using the latest data from the American space shuttle made available to the public just a few weeks before, and data from other NASA satellites. He took notes but he asked no questions, and eventually I fathomed out his identity – he wasn't a journalist at all, but Bill Woods, advisor to Condoleezza Rice and White House mapmaker at the Department of State in Washington. He had come to Darfur personally to take stock of the situation.

The next few days flew by. We were tormented by hunger and thirst as our food and water supply dwindled to virtually nothing. There were unexploded mortar shells amid the endless sand and rocks on our path. We made our way from one site to another with no visibility, in a fog of fine sand that penetrated our lungs and our computers, yet we maintained the daily quota of GPS points and geological samples. We had to find water, and fast. The refugee camps in Touloum and Iridimi, like the Oure Cassoni camp at the Sudanese border, had been hastily erected, and water was trucked in constantly, as no one knew where to find water underground. Yet the truck convoys were increasingly vulnerable to attack, their cargo stolen by rebels and bandits of every stripe who roamed the region and helped themselves, adding to the instability throughout the area.

Every day on the track we encountered despairing people and animals, awaiting death in resigned silence. Countless groups of refugees roaming the desert like haggard ghosts, their eyes begging for help. There were groups of children without parents, and groups of adults with no children, and always the stench of rotting bodies wherever we went. My driver swerved erratically to avoid shells and dust-covered carcasses on the track. Were the bodies human or animal? The animals died in their hundreds around dried-up wells, poisoning what little water remained in the process.

Death was everywhere. The atmosphere was surreal – horrifying, overwhelming, and yet simultaneously still, silent. One of the most harrowing aspects was seeing belongings sitting abandoned in the sand – desperately sad reminders of life. A bra, a shoe, an item of clothing. One day, trawling through the various bits of debris, I saw an image that wrenched my gut: it was a well-designed plastic suitcase, brand new, sitting upright on a sand dune. Who was the owner, and where were they? It conjured up conflicting thoughts – the suitcase itself was such a familiar image with its associations of travel, change, new experiences; it suddenly seemed European, comforting even. And yet it existed because of a heartbreaking mass exodus that was costing countless lives.

Gradually, our mapping work began to take form as the GPS points matched and confirmed our calculations. We were right; I watched the water surge from the wells we drilled near to the camps, overflowing onto the ground, and tears of joy streamed down my face.

It was a moment of intense happiness. We had come a long way in helping to secure success and victory in the face of pervasive and overwhelming death. The cynicism that surrounded our endeavours had been proved wrong; the madness of the last few weeks had paid off.

There is nothing more profound, more extraordinary, than providing drinkable water for those in desperate need. Everything we'd seen, everything we'd experienced, was now worthwhile.

Discovering oil gave me pride. Discovering water gave me light.

After this success, more were to follow. I was the first to identify the new site of Dalal-Gaga in December 2004, near a dry riverbed in eastern Chad, 60 kilometres east of Abéché. I advised running several test-drills for water before proposing this site as a new refugee camp. The description of this site was a significant chapter in my very first reports on eastern Chad, sent to the UNHCR in January 2005. The camp became a sort of focus point to calibrate all the other sites and rank them based on their importance.

Oxfam drilled according to our mapping and found three positive results; the water had a high flow rate and it was announced that the opening of a new Gaga refugee camp would take place in February 2005. In the meantime, some forty-five kilometres west of Adré, the camp of Farchana planned for 14,000 refugees based on its potential for water was now filled with nearly 20,000 refugees. The revolt that took place was totally foreseeable – shortages of food and water are commonly the basis of riots, and, of course, the place was hugely overcrowded. There were several casualties, including two UNHCR representatives, but it took the deaths of members of the UN delegation to convince the UNHCR that something must be done.

Our timely discovery of the Dalal-Gaga camp took the pressure off the failing, overcrowded Farchana camp and ended the riots. In just a few months, our drilling success rate had risen from 30 to 89 per cent. However, in order to gather all this data, I needed information from the drillers, and they resented my asking for data on their successes and failures – they perceived it as interference. They were paid per metre drilled, not per "quantity of water discovered", and saw no point in my statistics, unless it was to criticise their drilling – surely the only reason I would be interested in snooping on their work was to tell them what they were doing wrong. It'd be safe to say that I didn't make many friends in Darfur, especially with the NGOs.

They wanted no one intruding on their sovereign and exclusive power, and yet they were curious to know more about the advancement and success of our work. That was the most difficult and frustrating part of my experience.

I insisted on basic principles, on the collection of data; they knew, because of my findings, that I was witnessing them drilling in places that would inevitably result in failure. There was a certain imperviousness to their plans; they were insistent on continuing with the status quo, and I couldn't help cynically concluding that the fact that they were working with donor funds was part of the problem – money that comes so easily has no value, and I wondered on several occasions if this was what I was witnessing here. Statistics were important for our backers, who sought to save as much time and money as possible in order to save lives, but these financial partners were not on site to see for themselves the downward slide in people's conduct, and so they could not make changes that would improve things.

It was a source of frustration for us.

9: FIRST REWARDS IN THE UNITED STATES

Bill Woods, the White House advisor to Condoleezza Rice who I had met by chance in the dirt along the Sudanese border, invited me to the Department of State office in Washington a few weeks after my return from Darfur. He had been impressed by the success achieved with our technique, and by our operational vision. We greeted each other enthusiastically and affectionately, laughing together at finding ourselves in a comfortable office in the US Department of State when our last encounter had been in such different circumstances. I was, however, taken aback by how pale and thin he was, visibly in poor health. He made two announcements: that he had an inoperable brain tumour, and that the US Department of State wanted to fund our work to do the same kind of study we'd already done in Chad, and in Darfur in Sudan. This would mean covering 250,000 square kilometres, a surface area equivalent to half of France.

Prior to releasing the funds, however, the geological research and study bureau at the US Geological Survey (USGS) in Reston, Virginia, had to approve the system I had invented. I was hit doubly hard. Bill Woods had just announced very calmly that he was about to die, and he had informed me that we were going to be able to make the connection between Chad and Sudan. Was he leaving behind a legacy? That is how I perceived it when

I learned three weeks later that he had passed away. Bill Woods was and remains my primary mentor in this battle for water.

On his recommendation, in October 2004, the WATEX™ technology I had developed was validated by the hydrogeologists at the USGS research centre under Dr Saud Amer. Dr Amer is an eminent specialist in the use of remote detection climatic-related crisis management, and was at this time the scientific advisor to the White House on water issues for Africa, the Middle East and Asia. Dr Amer made a strong first impression on me. He is a tall, distinguished, good-natured man with an Omar Sharif type of moustache and bushy eyebrows. His face is serious and thoughtful, with a wide brow and empathetic smile.

A Christian Arab from Iraq, he had fled his country in the 1970s when it was under the control of Saddam Hussein. In the United States he completed a PhD in South Dakota and participated in the Landsat programme of Earth observation in NASA laboratories, and had eventually become a US citizen. Dr Amer had become one of the few world experts in radar applied to Earth surveillance; I couldn't think of anyone more qualified to assess the results we'd found in the Congo jungle and the Chad desert.

It has proved to be a fortuitous alliance – since our first meeting, he has come on site with me for all my projects, in collaboration with the Department of State, to some of the most challenging places on Earth, including Sudan, Afghanistan, Somalia, Ethiopia, Kenya, Kurdistan and Iraq. He has often acted as my guardian angel. Once, in February 2013, as I was collecting fossils from a dry riverbed in Somalia, he literally lifted me up and out by my braces, thus barely avoiding an ambush and kidnapping by Al-Shabaab militia fighting nearby. Dr Amer's nervousness was understandable given the summary execution of eighty Ethiopian civil servants just two weeks earlier. This happened some twenty kilometres away, along with the kidnapping of sixty others, most of whom would never see their families again.

This new, ambitious project of mapping Darfur started a new battle for water. In my offices in France, we were glued to our computers. We spent entire days, nights and weekends processing billions of pixels and making calculations to geographically reference them in longitudes and latitudes. We assembled, interpreted and printed out hundreds of satellite images. Within six months we had the overall grid of the underground landscape in Darfur: fossil rivers flowing with water, supplied by the Djebel Mara volcano, itself a veritable water tower that peaks at an altitude of 3,000 metres above hostile, arid plains ravaged by war and sun.

In early 2006, my team produced unbelievable results; we were absolutely astonished at what we'd discovered. From what we could see on our computer screens, there was clearly enough water for several million people. It was enough to stop the terrible war, to end the systematic destruction of the landscape, to rebuild the farming economy, feed the livestock and restore hope and dignity to an entire population. There was *much* more water in Sudan than in eastern Chad where I had achieved my first successes; the extent of what we'd discovered was almost impossible to fathom, even for us.

People were continuing to die of thirst in Darfur, for an obvious reason: this water was not visible on the surface. It percolated underground at a depth of 30 to 60 metres, in vast buried corridors dug out by ancient fossil rivers that had been flowing 2,000 to 3,000 years ago. We had a most vital question to answer – is this water replenishable? It seemed that the answer was yes; the radar images clearly showed that this buried radial network found its source on the sides of the volcano, which receives two metres of rainfall per year.

We had to go on site to verify and validate these results so as to convince the powers that be that we had really found something of significance here. But where could we go, how and with whom? This was not Chad, where we had protection from the army. This was Sudan, a war zone. We certainly could not count on any military assistance – its government was behind the genocide of its own people. I knew there had to be an answer to my questions, and I suspected it would come from Dr Saud Amer.

10: THE FRONT LINE IN DARFUR IS EXPANDING

With the assistance of UNESCO and the US Department of State, Dr Saud Amer organised in October 2006 a field trip and a training seminar in Khartoum to present the results of our mapping study to forty NGOs working in Darfur, primarily under the aegis of UNICEF. Due to the Sudanese government's involvement in the genocide, I could not provide the expedition with my water maps. We took the utmost care to protect the maps from the Arab horsemen who were decimating Darfur. They would have acquired all the land, which would have been easy to irrigate, and eliminated the indigenous people.

During our survey, Arab horsemen had already frightened me on several occasions, most memorably when I was driving to another training seminar, at the University of Dilling in Kordofan. We broke down at dusk in the midst of arid plains where South Sudan militia tanks were facing those of the northern forces; it was a dusty and hostile desert, scattered with stunted thornbushes – a daunting, sinisterly atmospheric place. Wolves lurked in the shadows, and howling winds swept across the sands. We were keen to get through the landscape as quickly as possible – instead, to our dismay, our vehicle suddenly became uncontrollable. We quickly deduced that the steering column in the car had broken on the rough roads – this was a serious breakdown, not something we could easily fix ourselves. We called for help by satellite radio, and I vainly attempted to reach Dr Amer in Washington simply to inform him of my whereabouts. I did not want to lose the scientific harvest I had reaped on the wells in the region, and I couldn't provide anyone with the coordinates of the site where I was to hide them. Desperate, I hastily dug a hole in the sand and buried all my notebooks and geophysical equipment, while awaiting hypothetical help. There was no way I was going to let the northern militia forces get hold of my maps, but I had made many unsuccessful calls to UNESCO in Khartoum, and now my satellite telephone batteries were dead. We were truly cut off from the world.

Night deepened; shots were being fired in the distance, and I no longer felt optimistic about the prospect of help arriving. The miserable possibility that we would eventually be discovered, but not by those who wished us well, was weighing heavily on my mind as we stared out at the moonless desert, listening to gunfire from somewhere nearby. And yet, to my amazement and intense relief, help turned up – a merchant from Dilling, travelling back home, came upon us purely by chance. It was approaching the middle of the night, and his wheezy old truck efficiently towed us to the city. Thanks to him, I was able to save, in extremis, all my results from the pillaging that would have inevitably occurred at daybreak . . . not to mention the risk to our own lives.

So my seminar took place as scheduled in Dilling, where I

worked with the drillers preparing the displaced-persons camps. Their mission was to drill the several thousand places I had identified in Darfur close to three big camps: El Fasher, Nyala and Al Geneina. Several million displaced people crowded into them in terrible conditions, surviving on the rations from the World Food Programme and the constant toing and froing of tank trucks bringing in water, at a cost of hundreds of millions of US dollars each year.

In Dilling I learned that three young Sudanese drillers, all in their early twenties, had just been killed by the refugees they had come to help. The drillers were working for UNICEF, and had been accused of poisoning the potable water cisterns when they were in the process of disinfecting the water storage tanks with chlorine. They were killed on the spot.

After the training seminar was over, I returned to France. From then on, I worked free of charge and unremittingly, with brave people who remained on site despite all the dangers. We worked side by side over the Internet, making use of a satellite link between the two camps during the weekends, when they returned from the field.

In the space of a year, from our offices in France, we successfully guided all the drilling, up to 60 metres deep, of nearly 1,700 wells. We regularly got flow rates of 20 to 30 cubic metres per hour; we had successfully secured the supply of water for all of these camps. Every Sunday, week after week, I provided a summary of all the drilling we did and prepared the answers to the many questions we got from the drillers, who often worked several hundred kilometres from each other. Darfur is a vast territory stretching from the eastern ranges of Chad to the western remote parts of Sudan over an area of 300,000 km^2.

UNICEF in Sudan used my WATEX™ maps for two years, drilling 1,700 wells for 3 million displaced people. Some camps themselves were moved to settle closer to the newly mapped buried water resources. The humanitarian crisis had been dealt with, but was the system I set up effective? I had to find a neutral partner to be in email contact with El Fasher, Nyala and Al Geneina and perform an objective scientific assessment.

I took advantage of the period to publish the first scientific report on the WATEX™ system and its application in Darfur, with the help of Dr Firoz Verjee.

Dr Verjee is a prince in the Ismailia tradition. He was at this time close to the Aga Khan, whose family came from Gujarat in India. He had spent his childhood in Kenya, then Uganda, where he fled persecution by Idi Amin, coming to England with his entire family; he eventually ended up settling in Vancouver, Canada. I met him for the first time in 2002 when he worked for RADARSAT International as director for Asia. He provided radar images to countries in crisis situations, images urgently requested by his clients during typhoons, earthquakes and tsunami events. These radar images enabled him to see, before anyone else, through the clouds, the extent of major climatic and geological events.

In order to jointly draft the article on Darfur for publication in a scientific journal on remote detection,* we met at George Washington University, in the ICDRM,† where he was a researcher and consultant to the USGS in the Department of Homeland Security. In addition to humanitarian concerns, Dr Verjee and Dr Amer also shared the same attitude of kindliness, their faces alive with intelligence and serenity whatever the circumstances in their tumultuous lives. I admired both men for all these qualities, which I did not have, and for all that I learned from them when we were able to meet. I was grateful for every opportunity we had to work together. I had the feeling we had known each other for aeons. Our meetings were utterly absorbing and further laid out the path for new actions, which we undertook with calm determination, our sights set on the long term. Saud and Firoz both played an important role in my ongoing involvement in emergency operations and conflicts prevention.

Two years later, after the intensive drilling campaign guided by our groundwater mapping of all of Darfur, UNICEF invited me to a seminar at the camp in El Fasher, former capital of the pro-slavery Sultanate of Darfur. The El Fasher camp already held

* Verjee, F. and Gachet, A. "Mapping Water Potential: The Use of WATEX to Support UNHCR Refugee Camp Operations in Eastern Chad," *GIS Development*, Vol. 10, No. 4 (April 2006).

† ICDRM, or the Institute for Crisis, Disaster, and Risk Management, is part of George Washington University in the United States.

50,000 displaced people. I was to explain our maps to all the NGOs, and put our drilling results into perspective.

Just as the plane was landing in Khartoum on 14 April 2008, Darfur rebels from the Justice and Equality Movement (JEM) and the Sudan Liberation Army (SLA) attacked the northern areas of the city just a few kilometres from the airport. The airport was in a state of panic – as soon as passengers cleared customs they ran to hide in the toilets and more secluded corridors of the airport. I skirted the combat zone and leapt into the last taxi to get to my hotel. There I remained trapped for three entire days under a strict curfew, with no food and no water.

Strangely enough, the hotel's Internet and phone connections seemed to work. It reminded me of my adventures in Moscow during the coup again Gorbachev.

I was able to reassure my family in France without any difficulty and to speak with UNICEF in Khartoum, but I hadn't taken into account the seriousness of the situation. I was completely isolated, trapped; the sound of gunfire in the city was constant, day and night. There was also the sound of combat helicopters bombarding the Omdurman Bridge, which connected Khartoum on the White Nile to Omdurman.

The next day, UNICEF informed me that all flights over Khartoum were cancelled. We were cut off from the rest of the world and prohibited from leaving the city, where fighting was relentless. After three days, the shooting subsided following the execution of the commander of the JEM forces. Surviving rebel soldiers were no longer hunted down in the streets of Khartoum, and I was able to walk to UNICEF offices to work with the drillers who had urgently come from Darfur.

It was still a difficult and uneasy working environment. The atmosphere was stifling, the white-hot city invaded by sand in a dry tornado full of lightning. We worked three entire days inventorying all the data on all the wells drilled for the camps. It was the end of the dry season and temperatures in Khartoum reached 49°C. We were desperate for rainfall, and everyone had holed up in dust-laden houses, fearful of more gunshots.

There was good news, however, despite the difficult circumstances. Since 2006, the drillers had used my WATEX™ maps to drill 1,700 wells, with a 98 per cent success rate. Though obviously I had hoped for success on this scale, I was thrilled that we had accomplished it.

Darfur is immense. It covers 512,000 square kilometres, approximately the same area as four-fifths of France. Darfur has a high range of volcanoes similar to France's Massif Central, the Djebel Mara, which peaks at over 3,000 metres with extensive annual rainfall. These hills are essentially water towers. During the British mandate, they flourished with citrus trees, and fed the Fur people. Sadly, repeated wars, raids and burning ruined this country, and it's now abandoned to the hyenas and the bandits. Though 1,700 wells sounds like a lot, it is actually very few when looked at in the context of such a vast area – the equivalent of 1,700 wells for the water needs of the entire population of half of France. But these 1,700 wells in

Darfur were the difference between life and death for 3 million displaced people crowded into camps surrounded by immense arid lands.

After the peak of the Darfur crisis, the subject no longer attracted journalists. The same was true for the statistical work required to certify the effectiveness of the WATEX™ process, which seemed to be of no interest to the local authorities. I can only attest to these results based on in-house publications from the USGS, written by Saud Amer and UNICEF's consultant hydrologist at the time, Douglas Graham. Both encouraged the continuation of my work.

Despite our incredible success, I went back to France with no new contract. Budgets had already been allocated and water was no longer of much interest, as the Darfur camps had all been secured by the wells we had identified, which were now producing potable, renewable water. I was no longer needed. Once again, I had somehow proved myself to be a poor negotiator; though I knew our wells had made a huge difference to human life, commercially I felt like a failure.

11: DISILLUSION AND PAIN

"A pessimist sees the difficulty in every opportunity; an optimist sees the opportunity in every difficulty." Winston Churchill

During my presentations in Geneva a few weeks later, I sensed something was troubling my contacts at UNOSAT. I could not leave my feeling unspoken, and I confronted them about my unease. They said, "Alain, it bothers us that you are using a technology that we do not understand; that's a serious obstacle to working with you in the future."

Reading between the lines, this was clearly an indirect way of asking me for the codes behind the new WATEX™ system, for free.

Intellectual property belongs to my firm. More importantly, the codes themselves would not even have been of benefit to them in isolation – the UNOSAT staff were not geologists nor geophysicists, much less oil people, and in their hands the programme would quickly degenerate; it would be like offering a car to someone without a driving licence. I reminded them that reality was catching up with us, and quoted the statement made to the press by Craig Senders, head of UNHCR Operations in Chad and in Darfur, about WATEX™ in 2005: "It has saved us a lot of time and energy searching for water in an area twice the size of Switzerland." Their purely academic attitude, which didn't take into account the reality of our work on the ground, was a crushing disappointment to me; I was saddened that they balked at working with a private company, despite it being pro bono, and despite all the evidence of our efficiency and success. I had not felt such reticence with the scientists at the USGS or with the political people at the US Department of State. So why was I experiencing this reluctance in Europe?

Because of this attitude, I was most likely not going to be able to work with UNOSAT again. Their preconceived notions appeared to be more important than the facts, more important than the lives we had undertaken to save together back in February 2004. It was becoming clearer to me that, even from the very beginning of our collaboration, UNOSAT's aim and focus was not to look for and find water. The task of UNOSAT is to map, from satellite images, natural disasters, detailed layouts of camps and infrastructures, and to convert such data into Geographic Information Systems (GIS) – its focus did not extend beyond this remit. It was disappointing that we were parting ways, but now I had realised the ways in which their focus was limited, it did not break my heart. There was, however, a profound sense of frustration, a sense that our fundamental principles were being betrayed. I expected intelligent, sensitive people with scientific training like myself, and thought that we would create a community of ideas that would lead to brainstorming, anticipating and building a future with new tools. It struck me as a rather pessimistic outlook, one that lacked courage and imagination and saw the problems inherent in every opportunity. Is courage the cardinal virtue of optimistic people, who see opportunity in every difficulty? Personally, I am convinced that it is. That is why I equate giving up with cowardice, the biggest flaw a mind can have, for cowardice leads straight to abandonment, betrayal and lies, and lies often lead to slander. This experience reminded me of how my boss fled the Congo, leaving me and my family standing on the runway in Brazzaville . . . it also reminded me of the irony of the courageous and wise Central African Republic's president, Ange-Felix Patassé, who honoured his cook simply because he did not flee with all the government ministers during the revolts in Bangui Capital!

Still, water sourcing continued, and in April 2014, I returned from an assessment mission in Chad. The authorities there confirmed that millions of dollars had been swallowed up by international organisations in fruitless attempts to find water. It simply allowed them to parade about in deluxe four-wheel-drive vehicles on the streets of N'Djamena and churn out unreadable reports loaded with embellished UN semantics. Not a single deep borehole and no new discoveries were made, yet they still arrogantly produced all the so-called thematic maps – essentially cut-and-paste documents, sold at a high price but with no new information contained within them. This mental posture, which unfortunately characterises most NGOs, brings to mind a quote from Albert Einstein: "Theory is when you know everything, but nothing works; practice is when everything works but no one knows why."

Above and beyond these vicissitudes, it seemed as though

the tragedy in Darfur was akin to a dress rehearsal, foreshadowing events that would take place soon on the scale of the entire planet. The Darfur tragedy was like a cancer, growing and spreading and gnawing away at the world, where the needs, due to galloping demographics, are exacerbated by a shortage of resources. Water scarcity, lack of land and, in the end, an incapacity in the ways and means of survival, are the challenges of this century, which today we call "climate change": what a euphemism! The riots that took place at the Farchana camp were there to remind us, on a very small scale, of the dangers created by such water shortages. Will we ourselves one day have the same experience? The current conflicts in the Middle East, other than religion-based clashes, are they not the consequence of shortages? They are communities scarred and weakened by a crucial lack of drinking water, water to irrigate arid land and to better feed their populations. The anger and frustration are palpable, and destructive.

In Darfur, I witnessed how economies based on shortage and survival are toxic and destructive – evil, in fact. They generate and justify genocides such as those which occurred in Rwanda and Bangladesh, and the permanent conflicts experienced in Somalia, South Sudan, Syria and Yemen. This is not an exogenous evil, a random blip, or a flaw inherent to developing countries. This is not a problem for "them". It is something that can affect us all, sooner or later, if it is allowed to grow.

I have observed with great sadness how peace and prosperity continue to shrink year after year in Africa, the Middle East and Asia. It seems like such a waste – we all know there are solutions to all these challenges, and that intelligent, sensitive people avoid confrontation and conflict, yet we consistently see those willing to sacrifice human values for the sake of a career. It's so important to continue the fight – to keep going. To quote Einstein again: "The ideals which have always shone before me and filled me with joy are goodness, beauty and truth."

12: THE FALL OF A DANGEROUS MAN

In September 2008, I was invited to speak to the European Commission in Brussels about a new approach to reducing conflict over water. As I arrived, I was informed that I would be allotted just a few minutes for my presentation. Someone from the UNDP* took me aside and told me, with amazing self-assurance, that our work in Darfur and in Chad had created uneasiness in the Commission, for my discovery of water in a war zone could worsen the violence. To my amazement, it seemed that I was considered "dangerous": someone who could shake up the status quo and incite even worse conflict in a troubled area. I could not believe what I was hearing, standing in an air-conditioned, sterilised corridor of the EU: no one here had seen the overwhelming distress of the Darfur children; no one here had seen another human being dying of thirst.

I left this costly institution on the same morning I spoke, in disarray, upset by what I had heard, and with a dizzying doubt about the very nature of my activities. My faith in what we were doing remained, and yet now I found myself on the edge of a void, feared and distrusted by the very people I was trying to help. After so many years of effort, of risk-taking and of blind conviction in vital work I was sure would eventually be

* The UNDP is the United Nations' worldwide network for developments.

successful, it suddenly seemed possible that it might all have been for nothing. And was there any truth in what I was being told? Could my work finding water in the most dehydrated places on Earth be, counterintuitively, harmful?

I had to take their concerns seriously, if for no other reason than the fact that it is these institutions that set the budgets. It was clear that my work was not understood by these bureaucrats, and yet there was a far more compelling and emotionally intoxicating clarity in the eyes of those who pleaded with me in Darfur, the people for whom I'd successfully provided water, and who had survived thanks to our efforts. Facts are blind; global decision-makers who look at statistics from behind a desk should never ignore this truth. The discovery of water, it seemed, could actually disturb the leaders of a country, for such water could be conducive to an entire population of refugees settling there when in fact their presence is not desired.

Refugees rarely return to their own country; why disregard a situation that is almost certain to endure? Every one of the Darfur refugees stated that they no longer recognised Sudan as their country. Women had been raped, men had been killed, wells were poisoned, and huts and homes were burned – they no longer had a home. Sudan mistreated and disowned Darfur people because of their dark skin; there was no inclination among them to return to a place where they were not wanted, a place where they were persecuted.

Humanitarian institutions raise funds more easily for urgent issues than for development projects. Projects that involve hope and prosperity aren't received as well, ironically; they are less easy to sell than tears, blood, guts, famine and despair, the basic elements for publicity that effectively gets funds out of donors. Humanitarian organisations generally use more than three-quarters of funds for their own internal operations. Humanitarianism is a business like any other; you have to please your audience, and that means doing so has to take precedence over your own conscience. Of course, this is not true for everyone. At the United Nations, I have known dedicated and sincere individuals who have done outstanding work. Unfortunately, they are lost in the ambient realpolitik cynicism, and can only impose their point of view with such great caution that they end up forced into the mould after all.

My consternation was on a par with my naivety. I had no real weight when it came to this immense humanitarian debacle. What was I supposed to do? What take was I expected to have, intellectually, in the face of this cowardly abandon? I couldn't bear to stand by and witness the decline of the very people I'd worked years to help, those for whom I'd invented and developed tangible, practical, viable solutions that could end their terrible misfortune. Humanity so easily forgets the lessons from the past; it seemed my work had served no purpose after all, and I experienced a deep sense of despair.

With no financial resources I could not keep going very long; I had to provide for my family, and I had nothing other than my work as a mining and oil consultant. My family were supportive, but I needed to keep a roof over their heads. I had the simple and obvious duty of protection and love that any parent has. My choice was made for me, heartbreaking though it was: I had to give up working in water, which was leading me nowhere other than to economic ruin and frustration.

It seemed that my work for the humanitarian community, then, was no longer wanted, and for five long years I was ostracised. To continue my work – and to continue to earn a salary – I accepted a few months later a gold-prospecting mission in Cameroon, near Equatorial Guinea. Though still reeling from my experience in Brussels, I accepted this exploration mission with enthusiasm – I was jobless and disillusioned, and I needed a new challenge.

As an exploration geologist, I began to reassemble the continental plates from 200 million years ago, like slotting together a jigsaw puzzle. It appeared that Cameroon was attached to Brazil before South America separated from Africa into two distinct continents. The state of Pernambuco, in Brazil, had at this time been connected to Cameroon and Equatorial Guinea and it was known to be a rich, gold-bearing province. I deduced from this that Cameroon's Lolodorf area extended over the gold-bearing zones in Pernambuco before the continental split occurred. The mineral riches in the wealthy province of Pernambuco had paid for the reconstruction of Lisbon after the terrible earthquake of 1 November 1755, which had been followed by a tsunami.

The day after my arrival in Cameroon, in November 2008,

the team of local geologists who had come to assist me headed for a hill with dense forest growth that I had ascertained was the most propitious for procuring gold. The driver parked in the shade, and we sat there for a little while before, suddenly, a group of Pygmies came out of the forest and onto the track. This was the opportunity I had been waiting for. We approached them and tried to talk to them; one of them pulled a peanut-sized nugget of gold out of a sack, and offered to sell it to me.

It is a mistake to combine technical consultation missions with mercantile takings, and so I politely declined his offer, simply asking them if I could take a photograph of their beautiful find in exchange for a small amount of money, which they accepted. The main question I wanted to ask, however, which explained my presence there, like a truffle-hunting dog, was where this nugget had been extracted from.

After something of a palaver – long discussions and negotiations, and the promise of good pay – they agreed to guide my entire team to their prospecting area the following day. And so, guided by a satellite image acquired on the other side of the Atlantic, I discovered previously unidentified gold-bearing sites know only to the Pygmies; they were the now-famous gold-bearing geological structures that extended from the state of Pernambuco in Brazil. The Pygmies had set up modest operations in the forest, with shovels, pickaxes, sluice boxes and pans to sieve the gold-bearing gravel from the fluvial terrace. After a good rainfall they sometimes found a few nuggets in the stream bed. A thin stream of water sluiced the stepped boxes, separating grains of gold from grains of quartz. Fine rain penetrated the dense treetops, soaking our shirts, chilling anyone who was not active. Everyone reached for a shovel or a bucket to stay busy and keep warm. We did not stay long. Just seeing the site confirmed my intuition that this was the extension on the other side of the Atlantic of the gold fields found in Pernambuco, Brazil.

We followed the path back to our car, in the rain. This forest was home to many types of wild animals, and the Pygmies told me of their misadventures with the small elephants, which can charge at the last minute – not reassuring to hear, as our visibility was just a few metres. Apparently, however, the greatest danger came from snakes on the ground, camouflaged by dead leaves; again, they could take you by surprise, biting with no warning. The Pygmies had keen eyesight and often spotted such snakes, but in any case they had "the secret of the black stone", which protected them from snake venom. My questions made them smile; they knew, as I know, that danger is always where you least expect it.

While I was deep in my own thoughts about the potential danger from the wildlife of the forest, a violent shock jolted my body and threw me high into the air. It must have been over in seconds, but I can still recall the vivid excruciating pain: my heart in my throat, my eyes out of their sockets and my adrenaline wild; the horror of realising that I was about to die was briefly at the forefront of my brain. After being harshly catapulted upwards, I landed heavily on the soft soil and fainted.

When eventually I came to, I was lying on the ground with my head in the mud, in agonising pain, unable to stand. I came to discover that I had stepped on an antelope trap.

I had a broken pelvis and my femur had been ripped from my hip socket and then miraculously reattached itself. The pain was unbearable. The initial air-toss may not quite have killed me, but I was going to face a slow agony in this remote part of the rainforest – this became in my mind a certainty. I began to feel irrationally angry that I had been abandoned by a God I had never previously believed in – if he was going to make an appearance, I thought, now would be a good time.

It wasn't yet my swansong, however; I was saved by my Pygmy guides, who dragged me, half-conscious, all the way to the car, more than a kilometre away. The pain was astonishing, unimaginable; I had no medication at all for several days as we were cut off from the world, in the middle of primeval forest without so much as a first-aid kit. After several days in the forest, it took an additional day for the driver who was eventually procured to get me to the base camp; the rough ride intensified the pain.

I couldn't walk to the base camp toilet, or take a shower on my own; I couldn't even tuck the mosquito net into my cot. I had literally hit rock bottom and my morale began to collapse too – it was vitally important, I suddenly realised, for me to reach the nearest hospital.

My Pygmy guides were touched by my distress, and came one by one to my hut to see me. They took my measurements with a dry branch cut from a nearby tree and made some rather expertly crafted crutches using wood and the stuffing from some old couch in an abandoned hut. They even fashioned skid-proof patches for the bottoms of the crutches with rubber from an old bicycle tyre. I was touched that they seemed to like me – though perhaps, on reflection, they harboured some guilt at having laid the trap so close to the path. I was airlifted back to France a few days later.

My return home remains a bad memory. A week earlier I had left in the best of health and in relatively high spirits, and I was returning in an ambulance like an invalid. My unusual crutches brought fits of laughter from all my friends, but the humour was somewhat lost on me, as I was beginning to understand that I wouldn't walk for the next six months. My horizon was clouded and unclear; I needed a total hip replacement and many long months of recovery. During my convalescence, it seemed all of my hard-won contacts, all of my fought-for projects, had suddenly evaporated. My country home had to be sold, and a mortgage taken out on my family house; after everything had seemed so promising and so positive, I found myself, to my astonishment, on the brink of bankruptcy. This hellish scenario was worsened by the financial crisis, which had in any case put paid to many projects in late 2008.

I was an explorer forced into invalidity and confinement; my frustration and misery were immense, but eventually I began to take stock of the situation. So far, my solitary nature had driven me to go it alone, so that I could move unhindered towards my objective without distractions or compromises. But now, if I wanted to go further, I realised that I had to break away from this solitary way of working and establish alliances, build solid bridges with other partners who could help ensure the sustainability of the approach I had fine-tuned alone. How could I proceed with this goal of drawing in new people? My career had hit a dry spell thanks to my injury, and the idea that I would be an attractive proposition for investors seemed unlikely. My wife was by my side to keep my spirits up and to talk through ideas with me, and she came up with a specific way in which I could reorganise my activities with reliable partners. Still, for the time being it all seemed very abstract to me. My morale was at an absolute low when, in 2009, a call came through from a friend working for the UN in Iraq requesting my assistance in the wake of the severe drought that had struck Kurdistan and Syria. It seemed my life was about to change, and water was beckoning once more.

13: DROUGHT IMPACTING THE HORN OF AFRICA AND MIDDLE EAST, 2009–2011

After a long period of recovery confined to my bed, in July 2009 I was called by the minister of water of the Kurdish Regional Government (KRG) of Iraq as a matter of urgency. I summoned enough strength to make the crucial decision to travel to Kurdistan, flying from Frankfurt to the capital, Erbil. The purpose was to find a response to the water crisis that had struck the entire region. In Erbil, I was hit by waves of heat as soon as I left the air-conditioned airport. The security had been reinforced with armoured vehicles around the airport compound, and there was nervousness all around.

Unrest had been brewing in neighbouring Syria since the 2007 to 2009 drought. It was the worst long-lasting drought ever recorded in Syria and the greater Fertile Crescent, causing widespread crop failure, livestock mortality and a mass migration of farming families to urban centres. The situation was perilous all over the Middle East, and ultimately prompted the Syrian uprising in 2011. For Syria, a country marked by poor governance and unsuitable rainfed-based agriculture and groundwater policies, the drought had a cataclysmic effect, leading to the civil war that continues to this day.

The UNESCO programme officer was waiting for me beyond the gates; he had not been cleared to pick me up at the airport.

He apologised profusely and explained that the situation was currently extremely tense – even the scientific director was not authorised to leave the United Nations defensive camp in Erbil to come and get me. It was terribly hot and dry; during our brief discussion in the glare of the sun, the soles of my shoes had begun to melt on the asphalt. I passed through the gates and climbed into his car, and two armoured cars driven by armed peshmerga in uniform escorted us to the UN camp entrance.

The United Nations camp was surrounded by huge concrete blocks chained together and guarded by US Marines. I was searched by soldiers from the Fiji Islands; they were giant warriors, both the men and the women, tattooed from head to toe, rather menacing in appearance but extremely polite and gracious in manner. An Oxford-educated Texan welcomed me and summed up the situation. The conflict over oil, and simultaneous jihadist attacks at the Kurdistan border and in the Iraqi Turkmen zone of Kirkuk and Mosul, had ratcheted up the tension between the KRG and the central government in Baghdad.

The intense dry weather of the last few months had weakened the KRG and there were unmistakable signs that there would soon be terrorist attacks in Erbil. Kurdish leaders had been decimated a few years earlier in an attack that claimed 200 lives, and the peshmerga were not going to let that happen again.

I was somewhat relieved to learn that there was no room left for me in the containers of the UN living quarters, which were hugely hot under the relentless sun. I could easily find, outside the huge concrete blocks, a room downtown run by Yazidis, who would also, apparently, ensure my safety.

The Yazidis are one of the most ancient communities in Mesopotamia. Considered heretics and devil-worshippers by some Muslim groups due to their reverence for Tawûse Melek (the Peacock Angel), a central figure in their belief system, Yazidis have faced persecution and genocide for centuries. They owe their salvation only to the mountains of Kurdistan and Djebel Sinjar, and to occasional protection from the Kurds. The Yazidis have a specific set of beliefs that are not aligned with any particular god or prophet – they regard good and evil as living in everyone, and believe that it is the free will and responsibility of each of us to choose the path of light based on a basic concept: to lie or not to lie. They believe that our entire lives are a battle between light and darkness, and that in each of us light and the forces of good must come to the fore in order that truth may prevail over lies and slander. Their creed is that every free man or woman must be the arbiter and the guide of their own conduct, every day, and at every instant. This is why I had been attracted to the Yazidis since my earliest encounter with the Zagros Mountains, which I find to be a highly spiritual spot on a desolate part of the Earth.

The next day, an armoured vehicle picked me up at my Yazidi hotel and we travelled to the Ministry of Water Resources to meet the Kurdish minister, surrounded by his colleagues and the press. Speaking before television cameras, the minister announced that research into water had been a top priority in Kurdistan ever since Saddam Hussein had ordered the destruction of all the traditional *qanats*. These gently sloping underground channels, not unlike a kind of underground Roman aqueduct, were designed to transport water from the mountainsides and carry it to the villages scattered throughout the mountains. In destroying the *qanat* system, Saddam Hussein sought to bring the Kurdish rebellion to its knees. He forced the thirsty residents of the small, scattered villages to relocate to the big cities in the plain, where he could better control them. It's staggering to remember that between 3,200 and 5,000 recalcitrant Kurdish men, women and children were gassed to death in the Halabja chemical attack in 1988 in a remote part of the Zagros Mountains near Iran.

After this long presentation, I expected to be offered a project that would allow me to prove the efficacy of the WATEX™ algorithm, but then it was made clear that I had just been invited as a European expert to speak in front of the press, to bring some kind of political relief to the auditors and the politicians. I would have to come back a year later to meet once again with UNESCO to build a strategy for Iraq.

At this time, Baghdad was "the Devil's cauldron", an incredibly dangerous place, as proven by the suicide bombing that demolished the UN headquarters there in 2003, an attack that claimed the lives of twenty-three people and wounded almost three hundred. After this terrible atrocity masterminded by

Al Qaeda, all UN-related organisations retreated to Amman in Jordan.

Along with Jericho, Erbil is one of the most fascinating cities of the ancient world. The Erbil plain is best reached by flying over Turkey and the snowy peaks of the Taurus mountain range; the flight is awe-inspiring, and the landscape changes radically in the low Mesopotamian plains. The alignment of anticline traps (formed by the folding of rock strata into an arch-like shape), including the Kirkuk embayment, looks like whitish limestone hills. It sits atop a bleak and barren plain, where the sandy dust dries up everything in its path. The lazy waters of the Tigris, depleted by the Turkish dams built upstream, thread their way south in partially dried-up meanders that still irrigate a few scattered plots of land. In the midst of the plain stands the ancient Erbil Citadel, like a precious pearl on a deserted beach.

In ancient history, Erbil was known as Arbeles, and it is where Alexander the Great defeated King Darius in 331 BC. The rounded hill of the citadel, with walls approximately thirty metres high, still holds the city, which is around ten thousand years old and therefore one of the oldest cities of the ancient world. Arbeles withstood countless sieges, for it had buried water resources from secret underground channels called *kareez*, which supplied fresh water from the Zagros Mountains that tower over the landscape some twenty kilometres away. Enemies who laid siege to Erbil died of thirst, while the city had an abundant water supply. Alexander the Great had his summer residence in Arbeles and came here when the hot Mesopotamian plain made Babylon, despite its fountains and hanging gardens, unbearable.

This area – Kurdistan – which has been ravaged by more than a century of warfare and rebellion resembles Switzerland without the cows, without the trees. Saddam Hussein ordered all fruit trees cut down. He had the livestock slaughtered and destroyed the water-channelling systems, the *kareez* or *qanats*, that had been in use for several thousand years. The entire area needed to be rebuilt and reforested.

14: PROTECTED BY A MONGOLIAN BATTALION IN CHAD

Six months after my adventure in Kurdistan, in February 2010, an urgent telegram was radioed by the Chad Ministry of the Interior and Public Safety informing us of the start of rescue operations for 50,000 refugees from the Oure Cassoni camp at the Sudanese border. The wave of drought that started in 2008, reaching from the Horn of Africa to the Fertile Crescent, worsened through 2009 and eventually affected the entire Sahel region.

The telegram asking for refugee rescue operations read:

> *National Commission for the Reception and Reinsertion of Refugees STOP*
>
> *Within the scope moving the Oure Cassoni refugee camp STOP in the Haouach area STOP a technical crew STOP government STOP UNHCR STOP continues field study in the Amdjarass area STOP to prepare reception and settling of the refugees at the selected Haouach site STOP Please alert traditional administrative authorities STOP local population STOP to enhance success of the technical studies STOP to start camp relocation STOP and END.*

Suddenly here I was again in Chad with the UNHCR, working on preparing a new rescue plan for the refugees six years after my first mission to Iriba in 2004 – and I had only been here a week. Christian Guillot, the young, enthusiastic representative of the UNHCR, had alerted me several weeks earlier to the crisis taking place in the Oure Cassoni camp; located at the Sudanese border, it was being continually raided by Sudanese Arab horsemen, backed by Libyan militia, who threatened the lives of 50,000 refugees daily. Several young women had been kidnapped to be sold as slaves to rich landowners in remote oases in Sudan and Yemen. Working from my office in France, I highlighted a security zone near a site called Camp Haouach, which possibly had large buried water reserves for these 50,000 people.

I took off from Abéché in a small United Nations plane that landed in Iriba. At first, I didn't recognise the site or the laterite airstrip. A wide asphalt road now ran along a military base housing several hundred people. I discovered a defence camp with several dozen buildings surrounded by trenches and huge sandbags. A veritable bastion by the Sudanese border, it was a hive buzzing with armoured tanks, jeeps, cranes and trucks operated by the blue-helmeted UN peacekeepers from the MINURCAT.*

The village of Iriba had become a prosperous town, and the UNHCR offices were behind high white walls with a big blue gate in the centre of it; the town's streets were still dirt roads, but it was undoubtedly an established settlement.

* MINURCAT: United Nations Mission in the Central African Republic and Chad. Contributes to the protection of civilians, promotion of human rights, the rule of law and promotion of regional peace.

The sultan, Bakhit Hagar, seemed to have grown considerably richer with this sudden influx of boreholes in his gardens drilled by all the NGOs working to his remit. He reigned over his subjects with a strong hand, in parallel with local authorities. Here, no decision was made without Sultan Bakhit Hagar's approval. Hagar was a clever businessman whose fortune came from the water rights and hereditary land in his possession. His resilience during the great drought of 2004 had clearly been successful in Iriba; he now owned a bakery and was in charge of the market, now much more prosperous than the pitiful market from my previous visit, which had stocked only garlic and out-of-date tins of food.

Christian Guillot accompanied me to the military camp known as Sierra Base and updated me on security in the area. We would be housed at the camp and not in the town. It was forbidden to go out without protection from the DIS,* made up of Chadian militiamen. It was impossible to distinguish rebels from the military. This was a high-risk area 1,000 kilometres from the capital, in fact a long way from everything, despite having the appearance of a defence camp. Anybody could shoot anybody here – you could leave the camp only in organised convoys, protected by detachments of armed soldiers. He reminded me that, in terms of hostage-taking, we were a prized target – especially me, due to my successful work here on my last mission.

* *Détachement Intégré de Sécurité*: United Nations-supported security force in Chad.

He did not say anything more, for the UNHCR had taken no statistics in Chad since 2004, unlike the Sudanese side, where the success of the drilling I had undertaken in Darfur in 2008 had been established by UNICEF. It was impossible to determine whether this was incompetency or negligence; it struck me as a question of indifference from the various heads of mission, who were not particularly interested in water.

Christian Guillot dropped me off in front of my bunkhouse situated in an air-conditioned container, much more comfortable than the insect-infested hut I had occupied six years earlier. He gave me a week's worth of combat rations. I could even take a shower between six and eight every evening, tangible proof that the boreholes I had identified six years earlier were flourishing!

After Christian's swift reception, I went to camp headquarters to meet the Mongolian delegation commanded by Colonel Bayar Lomsgooi, Major Ba'Atar and his aide-de-camp Gantschukh, all from Ulan Bator. Their battalion of eighteen Mongolian MINURCAT peacekeepers would provide security during my trips to the future Ouadi Haouach refugee camp (to be named after the wadi that sporadically supplied water to the area), which I had identified with satellite images. Seeing the extent of the protection force, I had the feeling that safety in the region had gone seriously downhill since we were last here, and that our expedition was going to be a real challenge.

The next day, I presented Major Ba'Atar with the maps I had

made for the layout of the future Ouadi Haouach camp. He was a tall man, solid and serious, with a wide face lit up by a kind smile. His calm concentration was a good omen, and he lent an attentive ear. He had just arrived from Sierra Leone after a long stint in Iraq; his men had great respect for him and were utterly obedient. He was a worthy descendent of Genghis Khan's army and he led his troops with all the dignity of a tribal leader – he had taken them into the devil's cauldron, and never suffered a loss.

We were going to drive nearly 150 kilometres on dirt tracks and Major Ba'Atar used the map of the region I had prepared with my radar images. It proved accurate in the chaos of rock, dried riverbeds and dunes of dust and sand. We then continued to work on our cartography in Colonel Bayar's office; fortunately, Major Ba'Atar and Gantshukh spoke excellent English. I informed them of the discovery of a large aquifer in the primary sandstone at Ennedi and thought that the future refugee camp should be located as close to this zone as possible.

Ba'Atar's fingers raced across the map; the future Camp Haouach was 140 kilometres north-west of Oure Cassoni, and 110 kilometres north-east of Iriba. This meant a major expedition into hostile territory, with military assistance, armoured vehicles and drilling equipment, on roadless, trackless territory. Seeing Major Ba'Atar frown and scratch his forehead, I sensed an even bigger obstacle. He asked me to confirm the location for the future camp: right bank or left bank of the Ouadi Haouach? I indicated that it had to be the southern edge of the

Ennedi plateau. He looked troubled; he turned to me, and stated that the area was not only full of landmines, but it was located outside of the MINURCAT mandate, which was strictly limited to the Ouaddaï area. This double blow hit me hard. There was no water in the Ouaddaï for the refugees; our only hope was the Ennedi area, landmines or not.

It was impossible to venture into the area alone and without protection, especially as we needed to be there for a number of weeks to complete the work. A long and protracted series of communications then followed: we had to speak to the UNHCR representative, who then called the Abéché base, who then called the head MINURCAT officer in N'Djamena, and so on – it was a cascade of communication and bureaucracy that lasted all day, with a decision only made at the end of it. Christian Guillot, another official and I were to go to Fada, capital of Ennedi, to meet with the governor of Ennedi in person, Mr Ahmat Daddy, and get his authorisation to enter the area with his protection. Fada was a faraway palm grove in the middle of a jumble of yellow sandstone cliffs; our options were three days of difficult travel on land, or a two-hour flight on the rather rickety Caravan, a single-engine plane the UN used for short flights and landings on difficult airstrips. Neither filled me with joy, but we plumped for the plane.

Under a burning sun and clear skies, we took off from Iriba the next day, flying over Camp Sierra and then the village, dotted with the green squares of the sultan's well-irrigated gardens. We veered north eventually and found ourselves over the dried-up Ouadi Haouach riverbed, which stretched east through the granite landscape. Further off, immense columns of sandstone looked like the pillars of the Egyptian temples as they towered above sparkling yellow sands.

The relics of dried-up riverbeds were evidence that just a few thousand years ago, water and life abounded in this mythical landscape. There is prolific rock art, engraved into the sides of the sandstone cliffs with scrapers, hand-axes and flint arrow-heads discovered when sandstorms shifted the soils. Flying over this area of immense beauty was a magical experience; looking down was an opportunity to explore the past for any-one who can read the signs that are there to be seen, like an open book on prehistory. The pages turned under my eyes as we flew, then landed at Fada, at the base of high cliffs that sug-gested new adventures.

It was the middle of the day by now, and scorching dry winds blasted us as we jumped from the plane. Violent gusts of sandy wind forced us to bend low to the ground. A van picked us up and took us to the Prefect's Palace for a brief but decisive meet-ing. Everyone huddled around maps and I described where we hoped to find significant aquifers in the Ennedi area; I thought it was possible we could house 50,000 Sudanese refugees from the Oure Cassoni camp. It was also possible that we could build new roads to open up this zone, which was going to be able to develop agriculture once the water was flowing. I finished my presentation by declaring that we should study the entire Enn-edi area, as it had fabulous prospects for deep water. We hastily finished the discussion; the roaring winds were increasing in intensity and, if they became any stronger, we wouldn't be able to take off.

Governor Ahmat Daddy, on whom our mission depended, expressed keen interest in our project and assured us that he would send the proper instructions to Abéché, allowing us to explore the Ennedi foothills as we had suggested. We literally crawled back to the plane and got away in the nick of time; the Caravan would not have stood up to the storm that was brewing. We took off like a feather and hung on for dear life while the plane climbed to an altitude that calmed us a little. This lightning-quick meeting had been efficient and successful, and we felt a sense of jubilation; it was the first step in addressing our issues, but it was certainly not the last, as the unfolding adventure revealed.

In the meantime, at Camp Sierra headquarters in Iriba, Major Ba'Atar had readied the reconnoitring expedition for Ouadi Haouach. He prepared his Mongolian battalion of eighteen soldiers and an armoured vehicle with a cannon to ensure security, as well as to transport weapons and munitions. A Russian KAMAZ truck carried the future camp's supplies: tents, cots, spools of barbed wire. Five jeeps were to transport his soldiers and MRE individual combat rations. He also brought in another convoy with a drilling rig that contained all the logistical equipment we'd need.

Two days later, in the pearl-white dawn, the heavy convoy slowly set off. One after another, the vehicles cleared the camp barricades to cross into a zone no one had ever driven before. A DIS truck led the way. On board the Chadian militiamen said they knew every step of the way to get to our destination on the edge of Ouadi Haouach. One hundred kilometres as the crow flies from the camp, through a jumble of granite boulders and sand dunes, we got lost.

At sunrise, we drove through the sleeping village of Iriba, rousing the emaciated dogs lying in the dust, and headed northwest in the direction of the village of Ourba, which consisted of a few wells and a hundred camels. Our column drove through herds of livestock, then the desert took over, gaunt and flat on either side of our dirt track. Thirty kilometres outside the village of Nouna, the track came to an abrupt halt. Huge granite boulders thrust skyward, and scree poured from a canyon, blocking our trail. We made a huge detour but hit a mass of rock. It was impossible to go any further, especially with the heavy APC, which struggled on this rough terrain. The column stopped behind the DIS truck, and the tension was palpable. A convoy of stationary vehicles is the perfect prey for rebels who know this terrain incredibly well. The DIS soldiers were having a heated conversation, waving their arms, pointing left and right. They had no idea if we should go east or west, but they did not want to admit they were lost.

An uneasy sense of their limited competency had begun to reveal itself in the skirmishes that had occurred on our outings in the previous days – these soldiers resembled Rambo in their weapons and swagger, but it seemed to be smoke and mirrors.

Most had come from southern Chad and knew very well they were totally outmatched by the warriors from the north, members of the Zaghawa and Goran tribes. These tribes were precisely who we were going to encounter sooner or later, and the soldiers were fizzing with barely contained panic. This was the perfect scenario for an ambush, and I felt a creeping sense of fear and dread.

I took out my navigation instruments; there were no further objections to me guiding the convoy. They had lost face, their silence speaking volumes. I had studied every detail of the area on the radar images that were now installed in my navigation. After conferring with Major Ba'Atar, we both took over the convoy in his white jeep with the two aides-de-camp, armed with their assault rifles. To make it easier for the armoured vehicle, we drove on sand and not on rock. The maps clearly showed the possible passages in the parched canyons. We had to get there before dark. The countdown was underway!

We had to find the entrance into the Haouach basin system, camouflaged amid this chaos of rock; we were effectively threading ourselves through the eye of a needle, and the accuracy of the images we'd mapped proved vital. Within ten minutes we were driving on sand and in the shade between two high granite walls. Since it was not the rainy season, we ran no risk of being swept away by roaring water in the wadi, but we did have to move fast to avoid being ambushed, as we were again an ideal target in this narrow file, even if we could not be seen from the plain. We also had access to potable water if we were to need it.

The radar images showed the presence of water less than five metres deep nearly everywhere, with wells spaced at regular distances, bringing up water for the flocks and herds. These river systems, dry on the surface but flowing with water at a certain depth, were a significant lifeline for all the nomadic people in the surrounding area.

Soon we reached the Ourba wells, known as the "Wells of 200 camels" because so many animals came to drink there, served by women with ropes and buckets who brought up water for all the goats and camels. Several wells had seemingly been dug out by hand in cracks in the granite, confirming our satellite observations. We then headed to the Dourouba wells in an area of sand and dunes cut off by the sharp beds of small rivers, which slowed our progress. We were now surrounded by scraggly acacia trees; at one point the heavy APC got stuck in soft sand by a dry river, bringing the entire column to a halt. We managed to haul it out using ropes.

Just as the sun was setting, we reached the point we had identified ahead of time on the map: a small, stony hill overlooking the Haouach wadi. It was completely devoid of vegetation – stark and empty – and so could serve as a stable, open area that would be easier to defend when the helicopter transport missions got underway.

The MonBatt (Mongolian Battalion) strategy went to work and erected the camp in less than thirty minutes. No haste, no improvisation, everyone knew just what had to be done. Canvas, rope and metal tubes were assembled to create a big dormitory. Barbed wire was unrolled all around, a machine gun in each corner, gunners taking their position for the night. Big green bags were filled with gravel and reinforced with steel rods, then placed inside the barbed-wire defence line as protection from bullets and shells. A generator was set up, and floodlights lit the approaches to the camp. Everything was thought through in this tried-and-tested military ballet.

A sentinel observed furtive movement beyond the lit area of our camp. Apparently, our actions were being closely monitored by hostile elements; our camp had not gone unnoticed. Who was out there in the dark? We were ready for anything, theoretically, though the reality of knowing we were watched was somewhat terrifying. Around 8 p.m. a delegation of villagers paid us a visit, after having called out to us from afar, their calls interpreted by the sole Arabic-speaking battalion member, a lieutenant-colonel from the Tunisian navy who was with us as UN observer and translator.

I had no idea there was a village in such an inhospitable area, but here it was. This village, Washéké, was too small to be visible on our image's resolution, smaller than a pixel. The village chief came to inform us of his disapproval of foreigners on his territory without his authorisation. We did not let him enter the camp and the discussion took place under the spotlights. Christian Guillot, who remained in Iriba and managed communications between us and Abéché, was supposed to have officially informed the sub-prefect of Iriba, and of course Sultan Bakhit Hagar, whose approval was vital to our being here now.

Left: Drilling team cooling off in a pond of drilling mud at the Lotikipi 1 exploration well.

Below: The first exploration well in Lotikipi, drilled down to a depth of 330 metres.

Left: Children dancing round a well in the Lodwar aquifer, discovered by RTI and drilled in July 2014. They had never before seen that much water.

Below: Children in Turkana, in the lush vegetable plantation made possible by the new wells.

Below: The author overlooking the Kawergosk camp housing refugees from Kobani in Syria, and from Bakhdida near Mosul in Iraq, in January 2015. A young Kurdish soldier played the mandolin while ten kilometres away the bombing started near Mosul.

Above: High dike on the Mosul Dam that stands at 113 metres above the downstream bed of the Tigris, with the hydroelectric power plant in the foreground.

Left: Nawzad Hadi Mawlood, the governor of Erbil, who gave me the passes required for the evaluation missions in zones that had been recovered by the peshmerga.

Above: Rabia hospital bombed out by the coalition forces, especially British fighter bombers.

Left: Military control by peshmerga on the road to Sinjar from Rojava.

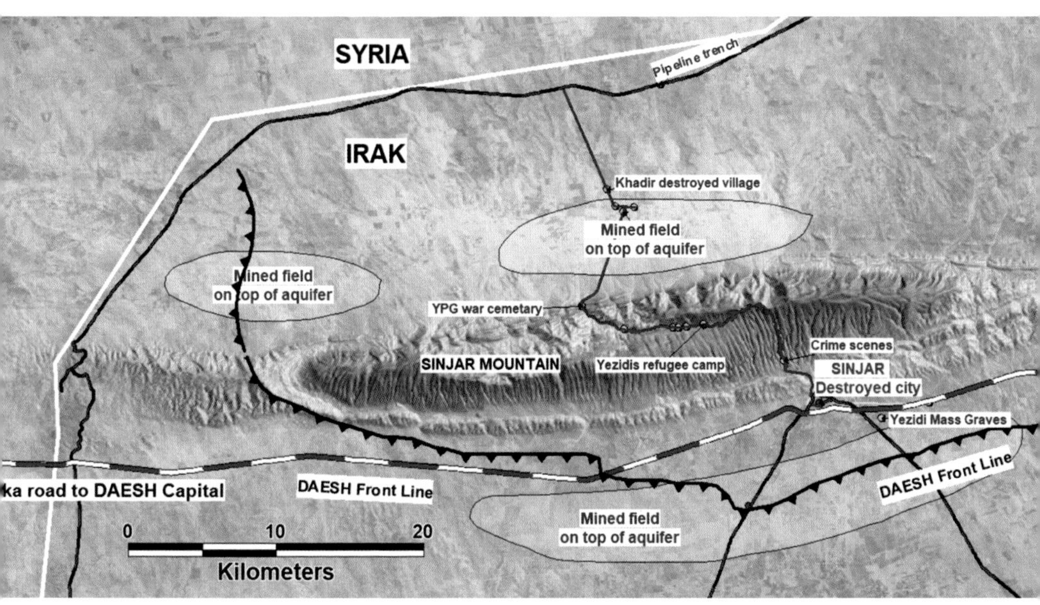

Satellite map showing the itinerary (blue line) taken by the author in the February 2016 mission. Aquifers are inside the blue polygons. The southern aquifer was occupied by Daesh on the front line and mined, as were the aquifers to the north of the sacred Yazidi mountain.

Sheikh Zarro, spiritual leader of the Yazidi community, mourns the extermination of his entire family, buried alive at the base of the mountain by Daesh.

Leader of the group of Yazidi fighters, a patriarch now driven by hatred. He is still strong enough to lead the resistance movement in the mountains with his one remaining son.

Torched and blasted vehicle at the entrance to the town of Sinjar.

Sign informs Daesh vandals, “Yazidis live here”.

Russian sign daubed on storefront by Chechen Islamist fighters.

Surprisingly, the impressive grain silos of Sinjar still tower over the plain, with slight scars at the top. The Sinjar plains were famous for their wheat crop.

Left: The Yazidi Peacock Angel "Melek Tawûse" on an armoured vehicle captured from the enemy on the front line, facing the village of Ranbusi controlled by Daesh.

On death road at the southern foothills of Djebel Sinjar, women and children were stripped of their clothing before being put to death or sold into slavery. February 2016. © Alain Gachet

Right: L to r: Astronaut Leroy Chiao, Gwynne Shotwell (president and CEO of SpaceX), the awardee Alain Gachet, and Daniel Lockney, the Technology Transfer Program executive at NASA Headquarters in Washington, D.C. at the time. © Alain Gachet

Left: Mala Bakhtiar (left) from the Politburo, executive secretary of the Patriotic Union of Kurdistan (PUK), whom I met in Sulaymaniyah on 24 February 2016, with my journalist and fixer Bakhtyar Haddad (right), killed in Mosul on 19 June 2017. We listened to his perspectives on the PUK.

Water Stress by Country: 2040

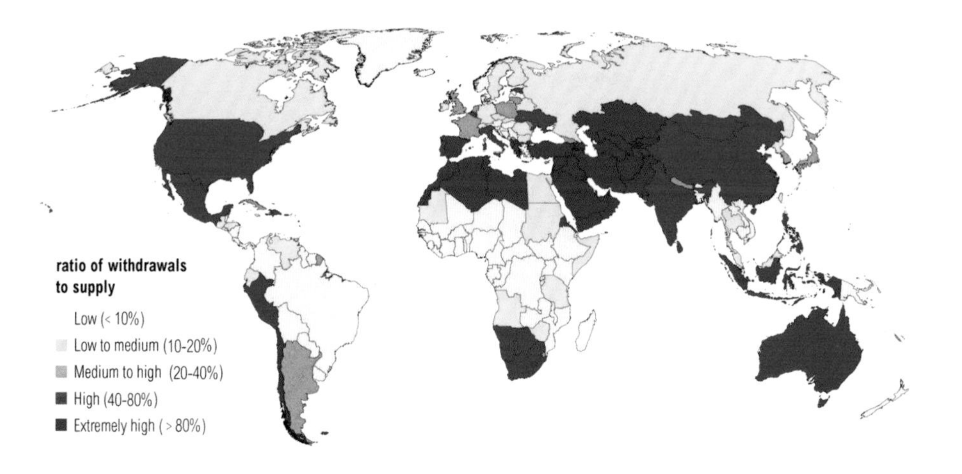

NOTE: Projections are based on a business-as-usual scenario using SSP2 and RCP8.5.

For more: ow.ly/RiWop

Above: Water Stress by Country in 2040.

Left: Monster dust storm on the tarmac at Niamey, Niger, June 2020.

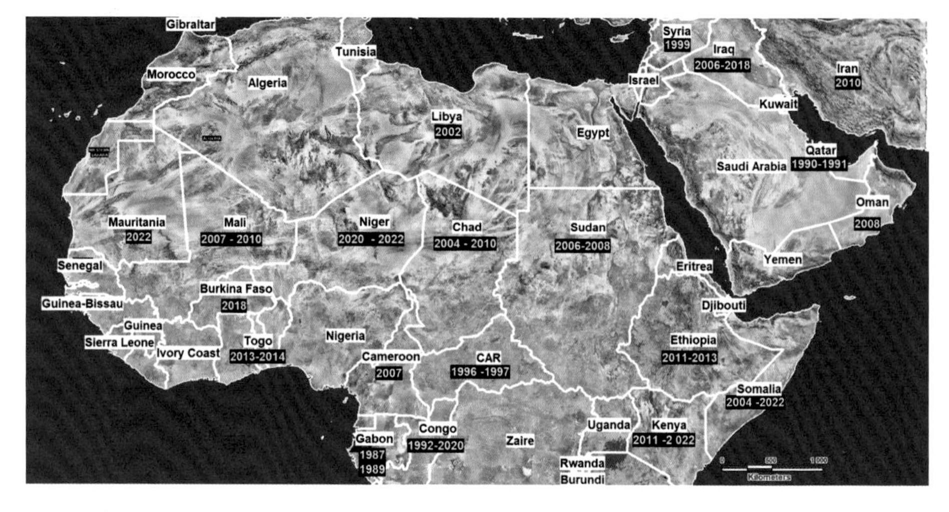

Map of RTI worldwide operations cited in this book.

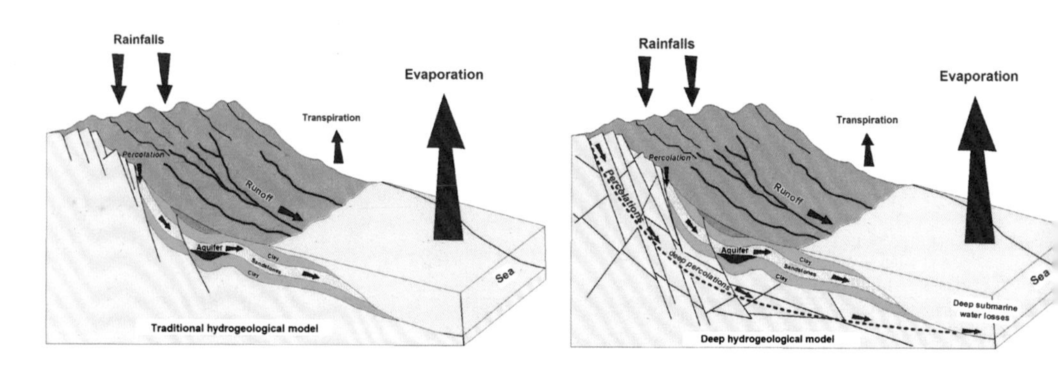

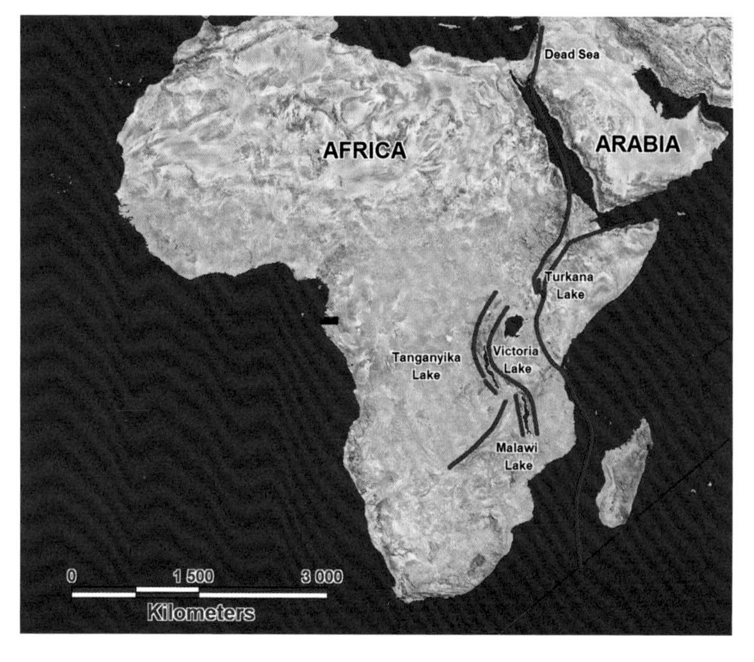

Left: East African rift fracture lines stretch more than 7,000 kilometers from South Africa to the Middle East.

However, our explanations of why we were there seemed to hold no weight with him, and the conversation grew heated. Several shadows darted around the camp outside the circle of the lights. Major Ba'Atar gave orders to his soldiers, who immediately took up their position and cocked their guns. The major turned towards the chief of the village delegation, who had grasped the power play at stake, and asked him to come back the next morning before 9 a.m. to discuss the issue in daylight. The night was quiet and windless; I slept on my cot under the open sky, next to the sides of the armoured vehicle for protection, like a baby rhinoceros snuggling against its mother. How was I ever going to be able to work in this hostile environment? Were we on the wrong track?

At dawn, all the vehicles were wet with dew. It was cold and the sky, a delicate shade of mauve, shimmered with the impending heat. In the distance, the currently dry meanders of the Haouach twisted and turned on the infinite horizon. The Haouach had been a significant river in prehistoric times, and I was certain we would find ancient sites of human occupation. A burning sun rose higher and there was not a single tree on this hill to shade us from its burning embrace all day long.

At around 9 a.m., the sentinels posted on top of the armoured vehicle gave an alert: a delegation requested admission into the camp. We refused entry but went to meet them a few metres outside the fencing, protected by the watching soldiers. Seated in a circle on the ground, discussion began at an advanced level of aggression. We listened in silence to the translation by the

Tunisian captain Mamdouh. The locals wanted no foreigners on site. We were a direct threat to their community. They wanted no outside aid and sought to make do with what they had. The colonel explained that we were there at the request of their very own government; we were there to look for water and open a camp for the many Sudanese refugees on the banks of the Haouach.

We tried to negotiate and reason with them, offering to drill a few boreholes for them in exchange for their help, or to build a passable track between their village and Iriba, to build a school for their children, to set up a medical office, to organise a vegetable-farming programme for their wives so as to better feed their families. But their gaze remained dark, their jaws tense, and their reply was firm and irrevocable. "We want no foreigners here on this land. We want nothing to do with your water, your schools *or* your hospitals. Our women are not your business, stay out! This is our land, not yours; get the hell out of here!" They were stubborn, firm in their conviction, and insolent – it left little room for hope.

In the meantime, several trucks appeared on the horizon, laden with men wearing long white djellabas, guns slung across their chests – they were vigilantes of all kinds, arriving from all the surrounding areas.

We understood, then, that there was nothing neutral about this discussion. The villagers themselves were simply proxy for the arriving rebels, who used them to get what they wanted. The guerrilla fighters reigned over immense areas through terror and ignorance, to obtain total control over all the different types of trafficking with neighbouring countries: drugs, arms, slaves, illegal workers to send to Europe. This was lucrative business in areas with no rule of law.

One of the rebel chiefs strode up to our group; our soldiers stood at the ready. A second rebel chief came silently forward, then a third left the group of trucks hidden behind the dunes and walked towards us. Each had the Thuraya satellite telephone so popular with warlords, but no weapons that we could see. We repeated our decree that we had come here not on our own but under orders from the government with a clear mission: drill boreholes, then leave. But it was clear that they believed we were there to take their land, an almost endearing interpretation of our presence given that there was not a single inch of fertile soil there, nothing but rocky treeless hills as far as the eye could see.

Any emaciated livestock there was surviving as best it could, making do with briny water. Despite the gravity of the situation, there was something rather poignant and comical about their ranting. If they could only see the rushing rivers that flow all year long in France, our vast green prairies, our Alpine forests, they would know immediately we were here only out of a sense of goodwill, not out of greed. Again, I felt a sense of despair as I contemplated how on earth our exploration could take place in such a hostile and unwelcoming environment.

Major Ba'Atar suddenly stood up to take a radio message from N'Djamena. It was the MINURCAT colonel backed by

Colonel Bayar Lomsgooi, who informed him that the prefect of Ouadi Haouar, Issaka Hassan Jogoï, who had come especially from Amdjarass at the express request of President Idriss Déby, was set to arrive any minute by helicopter. He was accompanied by Tidjani, the sub-prefect from Karoa. They had pulled out all the stops to show their support for us. They asked us to summon the entire population in the surrounding area for an emergency meeting.

Our adversaries heard this news and were electrified. The villagers seemed relieved to learn of this impromptu meeting, which would take the rebels' pressure off them. The rebels returned to their trucks parked behind the dunes, resigned but retaining their pride, for they had stood their ground. The discussion was over for the moment and the soldiers unloaded their guns.

While we waited for the arrival of the delegation, we decided to try to ingratiate ourselves with our new neighbours – we handed out bottles of water, and decided to pay an outrageous price to buy two sheep from the villagers for a shared barbecue. This timely diplomacy defused the tension. Faces relaxed and tongues loosened. The MonBatt soldiers unpacked their diesel-run field stoves – straight out of the Warsaw Pact. The stoves belched spirals of black smoke into the pure desert air, but anything was better than roasting the sheep over burning tyres on corrugated sheet metal, which I had often experienced in Africa.

We could hear the whirring as the two helicopters approached and landed one by one, blowing up pebbles and dust. The prefect Issaka Hassan Jogoï and sub-prefect Tidjani, escorted by the colonel from the Ouadi Haouar gendarmerie, made a stately entrance into the camp, followed by the UNHCR delegation headed by Christian Guillot, who stepped out of a Pakistani armed-forces helicopter.

The rebels and their trucks discreetly disappeared from behind the dunes. Under the tent, the village representatives saw that this display of political force had come to support our presence. Prefect Jogoï, a native of Amdjarass, a remote town in the north-east corner of Chad, had been well educated in Toronto as a legal expert and emanated the authority inherent to his warrior tribe. To the gathering in front of us he described me as a "water discoverer" and stated that "anyone who would dare touch even so much as a hair on my head" would deal directly with him. The same message was repeated and reenforced by sub-prefect Tidjani and by the colonel from the gendarmerie; each in turn stressed to the worried representatives the importance of our work there. We then unfolded the operations map and looked in detail at my plan of action.

We had to operate on the boundary of three counties: Ouadi Haouar, Ouadi Fira and Ennedi. We succeeded in obtaining three authorisations: from Prefect Jogoï for Ouadi Haouar, from Sultan Bakhit Hagar for Ouadi Fira, and from Ahmat Daddy, whom we had visited in the oasis of Fada, for Ennedi. We were legally prepared to do our work and, at last, everything seemed to be in order; alas, it was not to be. Major Ba'Atar received radio confirmation in the morning that the MINURCAT refused to widen the scope of the mandate of the Mongolian Battalion to Ennedi. This was precisely the area where I had identified the biggest potential aquifer to transfer the future Oure Cassoni refugee camp. I was devastated to see all this diplomacy and negotiation and red tape come to nothing. Inside the tent, where it was over 45° Celsius, I felt suffocated – by the heat, and by the situation.

The governor asked me to refocus my operations on Ouadi Fira to help the populations in the village of Waséké by drilling a well in their area. But my maps had shown that the residual potential of the aquifers in this area was very limited. It would be very risky and could well end in failure, which would be a disaster in this environment of crackling political tension, but the governor was insistent. Furthermore, my main objective was building 50,000 shelters for refugees in Ennedi. Lieutenant-Colonel Mamdouh came up with an unexpected answer. He offered to follow me to Ennedi, along with an aide-de-camp, as an unarmed UN observer. What was the point of UN observers in the event of violent confrontation with the rebels or the feared Goran herder-warriors? The answer was clear and unequivocal: the UN observer was there only to be a witness in the event of my kidnapping, and to give detailed information to UN headquarters. They wouldn't be able to help me, or intervene, but they could report what had taken place. Not exactly reassuring . . .

What purpose had it served to send out this whole convoy, and the seventeen Mongolian soldiers, to this southern bank of a dried-out river? Was it just to learn that a few steps further north, in Ennedi, they would be powerless if rebels were to take me hostage? I chomped at the bit for two entire days, and let two boreholes be drilled for the inhabitants at Waséké, even though I knew they would be dry. After this unsurprising failure I took off to explore further west, along the northern side of the Haouach. The two unarmed observers followed me around like shadows, hands in their pockets, on a Sunday stroll. I was the only one at risk. Since I could not use them as bodyguards I turned them into students, telling them about the geology we were seeing, under the murky gaze of my "official" bodyguard and a gaggle of three other gunslingers, heavily armed louts with twitchy trigger fingers.

We entered grandiose landscapes of breathtaking beauty in a jumble of sandstone and pink and grey granite boulders, studded with prehistoric sites and skilfully crafted stone huts. A few thousand years ago, an entire civilisation flourished here on the banks of this river that has since vanished into the sands, and yet I strongly sensed that there remained huge reserves of buried water not far underground.

Tchatorum, the Chadian security officer, was driving behind us when I stopped to take a few measurements with my ground radar equipment, a two-metre-long red antenna connected to a two-way power pack that looked vaguely like a pair of skis. The bodyguard jumped from his vehicle, determined to stop me getting my equipment out of the car. That was the last straw. I was refused decent armed protection and yet the little I had was keeping me from doing my job. Just what was going on here? Our conversation grew heated when I saw that he was literally trembling with fright; we had just seen a group of Goran herders and we were on the brink of entering enemy territory. But surely that was what he was there for?

I took out my geophysics instruments, knowing full well he could report on what he considered my irresponsible conduct, and told him I didn't care; the water I was looking for was not for me but for him and his country. I reminded him that his role was to ensure my safety, and to stand ready at the main protection points with his three other henchmen to cover me if need be. It was clear that he was afraid, but he was determined not to lose face.

There were three hours of daylight remaining and we had to think about getting back to camp. But there was still one last point on my radar images I wanted to see, a point that showed some actual moisture in this dry, Mars-like setting, just a few kilometres off the track we would use for our return; we arrived at it in a cloud of dust, the sun blinding us. Like a shimmering desert vision, or something out of *The Thousand and One Nights*, there suddenly appeared before us hundreds of camels and sheep, lit up like silhouettes against the sun on a rocky hill: clearly we had found a well.

"Gorans! Gorans!" shouted our gun-toting guardian angels. As we approached, we discovered Gorans indeed, but not quite in the form we were expecting – they were children, girls, surrounded by older men. One of the girls rode on a camel, and trotted back and forth to pull the rope lifting the goatskin out of a 50-metre-deep well; she had been working since dawn, she told us as she sat aloft in the wind.

Her tiny form balanced on top of her camel proudly; she had seen us approaching and wanted us to witness how she provided the water for this immense herd. She was only about ten years old, and not veiled. The old men poured the water from the goatskins into grooves that channelled the water to masonry drinking troughs dug into the ground. The other young girls channelled the goats on one side and the camels on the other, establishing a drinking order.

It seemed extraordinary that the well existed – it dated from colonial administration times, the 1950s most likely, and seemed to be working efficiently. My indications were accurate: under this stony plateau lay vast reserves of water that could help the refugees transferred here. And yet, one could not disturb the tribes, who maintained a delicate equilibrium in the immediate area – the refugees would have to go further afield. The elders at the well firmly ordered us to leave the site.

It was painful to see this young girl working so hard, at such a young age, for the good of her tribe. I pondered whether one day she would be able to go to school, whether she would ever read or write. It seemed desperately sad that her destiny would be limited to pulling a rope from the top of a camel – the life of a slave, effectively, except with the illusion of a sort of freedom perpetuated by the wide-open expanse in which she worked.

15: A SURPRISING OUTCOME IN ETHIOPIA, 2011

The Oscar-winning American documentary *An Inconvenient Truth* from 2006, about former US vice president Al Gore's campaign to highlight the effects of global warming, opened the international debate on climate change. At that time I was confronted with the Darfur crisis in Chad and Sudan, and I could witness how climate change was propelling me into groundwater exploration in Iraq, Somalia, Ethiopia and Kenya.

In July 2011, just one year after the expedition in northern Chad with the UNHCR and the Mongolian Battalion, I was contacted by UNESCO in Iraq. I imagined I would be invited to prepare a project to map the groundwater potential there. Syria was already engulfed in a spiral of violence that led to the ongoing civil war, and thousands of refugees had fled the country. In this context I was surprised not to be commissioned for an emergency programme for Iraq, but rather for a remote area in the eastern part of Ethiopia known as the Somali Region.

For our early, somewhat cautious approach, we selected a narrow strip of land in this remote part of Ethiopia along the Somaliland border. I had no choice but to accept this exotic and unexpected survey of the eastern part of the Great African Rift, and there were structural features that led me to think that some significant fracture lines might not only have an impact on the channelling of buried water, but possibly even hold large reserves of it.

The Somali Region of Ethiopia survey ended successfully: in less than four months I identified and mapped the extension of aquifers in several strategic places, and a new aquifer that was most likely very large. UNESCO could not and did not want to confirm it by drilling: the fear was that conflict along the Ethiopia/Somalia border would only be exacerbated by this discovery, and we were prevented from publishing this new mapping. As usual, I abided by and complied with prevailing geopolitical obligations.

The US Department of State was intrigued, and requested that I pursue the initial discovery and assess the actual size of the aquifer, under the management of Dr Amer from the USGS. Six months of intensive cartography ensued, both on and off site, with many trips through various parts of the region. Finally, we conducted with the IRC (International Rescue Committee, based in New York) thirty drillings to depths of between 100 and 700 metres – and to our delight, we confirmed that there was indeed a giant aquifer with replenishable water of excellent quality; the aquifer is just short of 250 kilometres in length, 20 kilometres wide, and 300 metres in depth. UNESCO never reported on this.

The very confidential results of this study have been declassified and can now be found on the USGS website, accessible to scientists since 2014. Given the current period of transborder tensions, these results have not been officially announced by the Ethiopian government, who fear that the presence of significant potable water would bring masses of Muslim Somali people into Ethiopia, allowing Al-Shabaab terrorists to infiltrate.

16: DISCOVERY IN THE TURKANA REGION

The great drought continued, all over the Horn of Africa and the Fertile Crescent. By the end of 2011, after Ethiopia, I was back in Iraq for the drought resilience programme. I was holed up in a bunker in the Green Zone in Baghdad, under constant threat of bombing. The drought then destabilised the whole region, especially Syria and Iraqi Kurdistan with their cohorts of climate refugees, infiltrated by the jihadists who organised the first demonstrations in the Syrian city of Daraa.

Once again I was contacted by UNESCO, and I expected to be asked to urgently map some other areas in the Middle East; instead, to my great surprise, I was commissioned to prepare a groundwater mapping programme for Turkana County, the most drought-affected area of northern Kenya. Violence had erupted between ethnic groups around wells and water sources, leading to a number of deaths.

In truth I had never heard of this region: where was it?

Turkana is the most remote province in northern Kenya, near South Sudan and southern Ethiopia, where the Omo River flows north of Lake Turkana. The western parts of the Turkana aquifers are recharged by the crystalline water flowing from the highlands of Uganda to the low-lying desert plains of the Turkana region. The north of Turkana between Sudan and Ethiopia was the tripoint fought over by the three riparian states. These tripoints around the world are conducive to smuggling and banditry, especially in the more remote areas of the planet. Tribes with automatic weapons supplied by the conflict in South Sudan have made violent incursions from one territory to another. Moreover, these tribal conflicts are exacerbated by the lack of water that threatens the lives of their herds, the only wealth nomads have.

Lake Turkana is saline and part of the Great Rift Valley. With an area of 6,405 square kilometres and a length of some 290 kilometres, it lies mainly in the territory of Kenya, with only its northern end in Ethiopia, and at the crossroads of the two branches of the Eastern and Western Rifts. This gives this entire area an exceptionally telluric* character, and its populations too. It was named Lake Rudolf by Sámuel Teleki and Ludwig von Höhnel in 1888, in honour of the Crown Prince of the Austro-Hungarian Empire, Rudolf of Habsburg-Lorraine. In 1975 it was renamed Lake Turkana. All these mysterious places evoked here in a few lines fascinated me from the very first expedition, because I knew nothing about this remote area of the world.

At the time of my first exploration in this part of northern Kenya in 2011, the Karamoja tribes, driven out of South Sudan by thirst, were again launching deadly raids against the Turkana

* Having a natural electric current flowing on and beneath the surface of the Earth.

tribes around the few water points still available, taking advantage of the situation to steal livestock to replace their own dead animals. Other tribes such as the Nyangatom and Dasaanech continued their raids, descending like a storm from Ethiopia's Omo Valley to seize the shores of Lake Rudolf in Kenya, pursuing fishermen with automatic weapons.

What could one do in such a context of violence and uncertainty? We had to dare to go, but I was faced with a blank page and I didn't know how to write the first line. I was simply mandated to practise my profession, to embark on an expedition of geological exploration on hostile territory whose uncertain cartography dated back to British colonisation.

As I was still in Baghdad, I prepared this mission holed up between sandbags at the bottom of my bunker. In the dark I traced the outline of my programme by following satellite images geographically recalibrated according to data from the American space shuttle. I had to be sure not to get lost on the tracks. I then tried to imagine, day after day, what awaited me once I got there. The presence of lions had been reported around the fortified villages, as well as giant crocodiles that would prevent us from camping on the shores of Lake Rudolf, and the thirsty and hostile bandits who would not fail to attack us along the disputed borderlands. But under shelling in Baghdad, all this seemed anecdotal, insignificant and distant; I had other emergencies to cope with for the moment.

It was almost impossible to anticipate events, and, when extracted from Baghdad by a military convoy to the international airport, I was still far from imagining all the dimensions of this adventure, with its twists and turns and the fascination. Fascination, certainly; if we acknowledge that these Turkana populations are the origin of the great migrations that have moved up the East African Rift towards Europe and Asia, they are the custodians of all our genetic heritage: we are all children of the East African Rift populations, and the Turkana are among our ancestors.

I was not expected to embark on this exploration alone; that would have been impossible given the situation on the ground. UNESCO offered me the protection of a military escort. I was able to leave the airport of Lodwar on a clear morning in January 2012 with my hammer and compass, and GPS installed on my armoured field computer.

My first encounter with the wilderness was the faces with sunken eyes pressed up against the hastily closed windows of my car. Skeletons in rags tried to keep pace with the vehicle, making begging gestures for food and water. We fled, driven by fear and by the horror of the scene, powerless to help despite our armed guards, who seemed unaffected, perhaps hardened to such sights. No one in Europe was talking about this; it was an invisible catastrophe, the result of five consecutive years of drought that had taken the lives of hundreds of thousands of cattle, and had had a devastating impact on the lives of 13 million people.

The geological maps drawn just after World War One by

the British army were not very accurate; you could get lost at every fork in the road and die of thirst. Luckily, with a good GPS linked to satellite images, we would know where we were driving as long as the car's cigarette lighter was working. Our batteries were dependent on unexpected constraints such as this. But, small mercies, a convoy from the Kenyan army was deployed to stay by my side throughout the Turkana exploration.

Along the track we came across a little girl whose parents had been murdered. No more than six years old, she sat in a hole that she'd dug with her bare hands in a dried-out riverbed, in a desperate search for water. She had a metal bowl, which she was using to help her scratch at the dirt, a plate, and two goats. It was incredibly painful to see her life hanging so delicately by a slender thread as she tried to collect some drops of water to give to her goats, who would in turn supply her with a bit of milk – she couldn't drink the water herself as it was too toxic for her stomach to withstand, and without water for her goats, there would be no milk for herself. Her situation was dire. She lifted her gaze, intrigued by the inexplicable presence of a man with light skin and white hair, and I held out my hand to help her out of the hole and got in there myself, to dig with a shovel. She stared at me with impassive eyes – perhaps hopeful, I wasn't sure. I was suddenly incredibly conscious of myself as a hopeless anomaly – what was I doing in this desolate area where jackals and hyenas prowled? I tried for a while, but I could dig little better than she in the hard bed of cemented pebbles, and was able to gather only a few drops of muddy water before handing back her tin bowl.

No rain had fallen for several months. The Karamoja warrior-herdsmen, observing from the highlands, moved down under the cover of darkness to ransack the herds and attack the Turkana villagers, their way of offsetting their own losses from the long drought period. These warriors used automatic weapons to kill several hundred Turkana herdsmen, to be sure of avoiding retaliation for a very long time. Several members of the young girl's community had been killed around the rare water holes – spears, leather shields and bows had been replaced by Kalashnikovs that in a single day had decimated her tribe, chasing any survivors – women, children and the elderly – far from the water.

The police were rare and thin on the ground in this immense territory; they did little more than hide in hastily dug holes in the bush, counting the dead after the battle to provide statistics to the civil servants in Nairobi. The wounded were left to die among the survivors who hid in the desert valleys of thorny locust trees, since they were hundreds of kilometres away from anything resembling a hospital. After many weeks in this vast area with my hammer and compass, tracing a network of trails detected by satellite image, which were barely visible from the ground, I concluded that we had to attempt to traverse them. It remained to be seen if our vehicles could negotiate these tracks, but we had to try, even if it meant turning back and retracing our steps after several days of travel. We had two off-road vehicles, which enabled us to make all kinds of detours to skirt the rockfalls that cut off the trail along the dry riverbeds, and helped us avoid the subsidence and landslides caused by the frequent earthquakes.

Turkana is the land of extremes: the morning coolness vanishes quickly, well before noon. During the day, we witnessed the beautiful vegetation and the bright-yellow dunes along the banks of Lake Turkana. But as noon approached, we'd be rapidly enveloped in a heat haze at temperatures that hit 40° Celsius, then gradually fell during the afternoon. We could only catch a breath of cool air after dark, under the starry skies. The area was unstable and savage, constantly disrupted by the seismic nature of the rift, which brought with it violent storms of rain or dust.

Before leaving, it dawned on me that there could be lions at our base camps. We carried no weapons, in keeping with the implicit rule in all these missions – our safety was entirely in the hands of the mobile army contingent that had accompanied us. As it turned out, the real threat was crocodiles. These monsters lurked near the villages, devouring dogs and, tragically, occasionally even children who had wandered too far from their huts.

The sandy banks around the lake were strewn with debris propelled by the strong winds that blow here. Amid carcasses of domesticated animals, there were fossils that were several million years old, torn from the bottom of the lake. Dinosaur vertebrae, giant fish bones, blue chalcedony stones of amazing beauty – countless treasures could be found on a good walk around the lake, whose alkaline waters tasted like baking soda. Fish thrive in the lake; I was surprised, however, to learn that the Turkana people refuse to fish in it, or to eat fish sourced from it, due to their superstitions. As a result, the lake was being harvested by seasonal fisherman who came from Ethiopia, finding the water buoyant and their catches bountiful.

At one point during this expedition, we had to go west of Lake Turkana, through a range of volcanic rock, to join up with the Italian St Patrick diocese mission for educating young nomadic girls. We entered the gorge lined with high walls of sandstone and crowned with basalt, and saw signs of several springs that had run dry long ago. As night fell, these narrow, deep canyons became increasingly sinister, and we walked with trepidation – these invisible hiding places seemed the perfect setting for an ambush scene in an adventure film. We finally reached the diocese safely in the pitch-black darkness, but the tension had persisted during the entire trip.

One afternoon, while I was sitting on one of the high terraces overlooking the river, I met a man whose legs had been amputated following a plane crash resulting from deliberate sabotage. His name was Dr Richard Leakey,* and he was an

* Dr Richard Leakey, in full Richard Erskine Frere Leakey (born 19 December 1944, Nairobi, Kenya; died 2 January 2022, near Nairobi), Kenyan anthropologist, conservationist and political figure who was responsible for extensive fossil finds related to human evolution.

eminent conservationist and ethno-palaeontologist who had made many enemies for having led anti-poaching groups all over Kenya. A pioneer in protecting endangered wildlife, Dr Leakey founded the Turkana Basin Institute, a palaeoanthropology research centre, on the Turkwel River at Lake Turkana.

The sandstone bluffs lined an oasis of greenery and palm trees and glowed pink and purple in the rays of the sun. In the gentle warmth of the late afternoon, we talked. As is clearly shown by the dried-up springs and ever-dropping level of the lake, climate change and its significant consequences is not a recent phenomenon and is not always and solely attributable to human activity. We took a walk, Leakey on his artificial legs. He told me that several lakes had vanished from the land, but that the populations remained. "They are condemned," he said, "to wander this land, abandoned by their gods, forever seeking their lost paradise." Just a few millennia ago, this area was verdant, traversed by flowing springs, abundant with antelopes, elephants, hippopotami and countless herds of long-horned ruminants, just like in the rock carvings and paintings in the central Sahara and Tassili n'Ajjer in southern Algeria.

As we continued to talk, he took me further back in time, to the era of Lucy, the small skeleton of an early female of the hominid species, discovered in 1974 in Hadar by an international team that included my highly respected friend, French professor Yves Coppens from the Collège de France. Lucy walked the Earth nearly 3.5 million years ago. At that time, the climate was very different and marked by swings between savannah forest and desert. "But," said Leakey, "humanity's very nature is to adapt to circumstances; it's up to us to find the answers to all of these changes. To paraphrase Coppens: 'Hard times and scarcity, not opulence or comfort, make men intelligent.'"

The fossils we had found during this mission allowed us to map the old lake level, using data extracted from the American space shuttle, though it took us a few weeks to put together all the pieces of the puzzle. Some freshwater fossils taken from the track around the lake showed me the former levels of Lake Turkana, 200 metres higher than its current level. This was clear proof that drought had gradually been making inroads over the last few thousand years.

A rift is the result of tectonic plates moving apart in the process of continental drift. It looks like a downfaulted depression, a narrow plain lined by high cliffs that follow a fault line. The Dead Sea Rift is the ultimate expression of the opening of the Red Sea, dividing Egypt from Sudan along the Arabian Peninsula coast. This rift is a central linear graben – a depressed block of the Earth's crust – framed by steep cliffs that extend for thousands of kilometres. The same fault line extends nearly 7,000 kilometres in East Africa, where the depression holds lakes such as Lakes Malawi, Rukwa and Tanganyika in Tanzania, Lake Kivu in Rwanda, Lakes Edward and Albert in Uganda, and Lake Turkana/Rudolf in Kenya.

This immense fault line in the Earth's crust crosses Ethiopia to the Red Sea, then goes northwards across the Gulf of Aqaba to the Syrian highlands after traversing the Dead Sea and the Sea of Galilee. I have been fascinated by rift formations since the age of fifteen, when I began studying Israel and the Jordan Valley. I understood the important role played by rifts in the circulation of water. Rifts through their long history have remained the ultimate resilience corridors against climate change, as they determine the courses of rivers. They gave rise to the development of mankind, providing water at all times. Great lakes are found in rifts; some are saline, such as the Dead Sea, some have immense reserves of fresh water, such as the African Great Lakes or the Sea of Galilee fed by the River Jordan.

Some cliff-lined rifts are obvious while others are much less visible, but both play a very important role. When filled with sediments, gravel, sand and mud, rifts disappear from view but their depths are there. These are the aquifers that remain to be discovered and which are deeply fascinating to me; there are rifts such as these in East Africa.

The wet surface of the sediments that cover these filled rifts gives an idea of what can be found underground, especially when traces of residual moisture are found along the length of the fault lines. This is the long-term work I have begun in Ethiopia, Somalia and Kenya, for here I have seen the immense potential: once again, a new branch of the future of mankind.

It is in this context that I mapped, in less than three months, the groundwater potential of all these rift territories. I recorded thousands of remarkable geological points there, before launching new exploration drilling on behalf of the UNHCR's refugee camps in Kakuma and Kalobeye at the gates of South Sudan.

These camps now host more than 300,000 refugees from Somalia, Uganda, Democratic Republic of the Congo, Ethiopia and South Sudan. It took some time to map underground water in these most deserted and hostile areas of the planet, where nothing suggested the slightest hope, and yet ... When the underground landscape was revealed in the light of my work, I quickly realised that these territories of thirst and despair were ultimately nothing more than territories rich in groundwater, but neglected by failing administrations. The nomadic pastoralists, our ancestors, are not voters; they cannot read or write and are of no interest to anyone except the NGOs, which, for various reasons, have each established their base camp in Lodwar. As a result, Lodwar, the administrative capital of Turkana County, a small dusty town an hour and a half's flight from Nairobi, has become, at the whim of successive humanitarian missions, the capital of poverty, attracting all the NGOs on the planet.

After strenuous work in the field, the WATEX™ images helped me to complete the big underground picture and revealed something that we couldn't even have begun to anticipate: five giant aquifers were hidden below this inhospitable countryside. Dr Richard Leakey had spoken to me of lakes that had vanished millennia ago. Were they here? How much water remained deep below the Earth's surface?

UNESCO had hired my company for purely academic purposes, to map the water potential of Kakuma refugee camp, but not necessarily to discover new water resources. UNESCO did not expect such a discovery – they did not want to approve it, or even dare consider it. If indeed there was water, such a fact would already have been reported, and yet no one had ever dared drill into these vast, hostile desert lands ... Moreover, such a discovery would inevitably require a huge commitment, and they did not want to engage their responsibilities in development: that was not in their mandate.

However, progressing this discovery was crucial: it was not a question of profit, nor was it some wild experiment with a tiny chance of success. The scientific data provided us with a high probability of finding deep water, greater than 90 per cent; the repercussions of this were immense. Of course, there was still the small possibility that no water would be found, and it was going to cost several hundreds of thousands of dollars to hire drillers. The work also had to take place in areas that were dangerous in more ways than one, in order to save communities and their flock. Nevertheless, the data was accurate and undeniable; I was able to convince my partners that there was no doubt, that the effort was worth pursuing. My certainty was based on the lessons learned during twenty years spent at the hard school of oil exploration. Exploration for deep water has, of course, many things in common with oil exploration – both require science initially, both also require physical exploration to validate theoretical concepts.

Otherwise, the concepts have no value; their authors would be robbed of their credibility.

So, the next step was to find a drilling company able to drill deeper than 400 metres below the Earth's surface, and test for the water potential in five aquifers scattered over a surface area roughly equivalent to half the size of Belgium. There were several responses to the calls for bids, all claiming that they were able to do the work; in each case, after some discussion, it was quite clear that the sheer scale of the project was beyond their capabilities. The size and condition of the compressors and their drilling equipment eliminated most of the companies from the start, obliging us to dig into a thick layer of lies and false promises; it took several weeks of searching, investigating and weeding out companies that didn't stand a chance before we were able to identify a firm that was qualified to help us confirm the presence of water in this unknown area.

In 2013, less than a year after the project had begun, the successful bidder's entire team started work on the surveying site in Lotikipi, under relentless sun. The drill mast was even taller than the termite mounds that stood in the middle of a vast plain and were covered with yellowed grasses, waving in the wind. Day in and day out, a tank truck had to bring in 15 tons of water from 40 kilometres away to the drilling site to create drilling fluid. Every evening, antelopes came and warily drank from the artificial pool dug to contain it. During the day, crew members swam there to relieve the sting of the sun, which beat down ruthlessly. Two shifts worked around the clock in the loud growling noise of the boring rods, which ground into the earth and were then picked back up by the mast. The compressor's vibration carried into the empty, rather eerie silence that surrounded us: there was not a single village, not a single tree closer than 50 kilometres in any direction. This desert land was totally uninhabited.

Suddenly, after this had been going on for a number of days and the drilling had reached a depth of 120 metres, the driver suddenly started shouting, "Water, water!" The pumps were spurting up streams of gravel in geysers of liquid – we had just drilled through a 20-metre reservoir, our very first victory. This was only the beginning, however – other, bigger aquifers awaited us deeper down, I was sure of it, and so the crew couldn't stop their work. We had to continue our quest, drilling deeper and deeper, as quickly as possible, in spite of their protestations.

Tension rose as the days went by. The drillers did not understand why I was not satisfied with our early discovery. A few weeks later, at a depth of 200 metres, there was a new eruption of gravel and water: an aquifer nearly 20 metres thick was discovered, and the drillers finally understood why we'd kept going. Their faces lit up; they were thrilled that we'd found what we were looking for, at last. Their moods were quickly dampened, however; we still weren't finished. I wanted to know what this well had to offer at a depth of 400 metres. They were visibly shocked – they clearly thought I was crazy.

At a depth of 280 metres, a third eruption spurted forth as we continued down to 330 metres, where we came upon a third aquifer nearly 80 metres thick. The compressor wheezed, and the strings of the drill pipe became too narrow for the lift pump to pass and conclude the production tests. We were not going to reach 400 metres, and I decided to stop the drilling. It was time to start the first tests on the three levels of aquifers, before the hole collapsed. The entire Turkana population came running to the drill site from all directions – women, children, even dogs all descended. How they knew, I will never discover, but it was an extraordinary sight to behold as the celebrations began – children and women danced in an abundance of water the like of which they had never witnessed.

In a final thrust into the Earth, the compressor roared and nearly choked to death in a cloud of smoke, pulling up a column of water and mud from a depth of 330 metres. Towards noon, with a deep gurgling noise that surged out of time immemorial, water gushed from the Lotikipi well before the incredulous eyes of impassive men seemingly lost in thought, draped in dignity, and women, dumbstruck and delighted.

Once the astonishment wore off, the men began by insulting their divinities for having let their forefathers die of thirst when there was – and had been for centuries – all this precious water underfoot. From every direction, lines of men, women, children and animals continued to converge upon this living fountain whose existence they had perceived from far away. This water would ensure their security – we had found reserves of hundreds of billions of cubic metres. This giant aquifer corresponded to the forecasts I had made using the new techniques

I had been experimenting with since 2004, during the Darfur crisis in Sudan.

The gigantic reservoir is part of the ancient sources of the Nile; this I had understood from the space shuttle data. The presence of Nile crocodiles on Lake Turkana further confirms this theory, for they do not belong to the family of Indian Ocean and Mozambique crocodiles.

My discovery opened a pathway of possibilities for these distressed populations. With such renewable reserves of water, every well could become and remain an oasis of prosperity, a space of reconciliation and peace. In desert topography, where scarce surface water causes disease and conflict, the very abundance of water should allow the opposite. Such a major discovery should result in sharing the increased resources, to the benefit of both nomads and sedentary populations. Several rivers dump yearly floodwaters fed by heavy rainfall onto the Ugandan and Kenyan highlands; the rivers course through the desert like so many lifelines before disappearing, absorbed into the ground. This was the lost water I had just brought into the light of day.

The children had never seen so much clear, pure and abundant freshwater that they could drink without limit. They played, laughed and danced under the impromptu shower. The women too danced spontaneously around me, the "white-haired wizard", thanking me with their songs. In the split second it took for the water to gush forth, they had understood that their lives had changed in a trice, and that they could now dream of an easier life, of prosperity.

Their dances were joyous and incredibly moving, a display of gratitude from among the most disadvantaged people on the planet. After ten years of effort, my dreams seemed finally to be realised and I felt hopeful for the future. The Turkana, though summarily identified as a nomadic people, had never left the harshness of the desert for elsewhere. Their nomadic wanderings took them from one point of the desert to another, moving with their herds, driven by the basic need to survive. The ceaseless search for water may have resulted from ancestral memories of a land that had once been prosperous. They did not need Moses to guide them to the Promised Land; they had their promised land underfoot, and their hereditary instincts told them to stay where they were.

As the first Lotikipi well began to flow, the community's spiritual leaders gathered right there before us. After the criticism of their gods, they turned upon the Kenyan government, who they said had never done a thing to mitigate their immeasurably desperate existence. Now, at last, they were going to recover their dignity. They were going to stop begging for food and water from hostile neighbours. They would stop risking their lives at remote and dangerous wells in South Sudan, where they could be massacred or have their cattle stolen. They were going to be able to settle and grow food to feed their families, there on their own ancestral lands. Their dream, too, was on its way to realisation, though more wells would have to be dug. Such were the words uttered in their prayers to their ancestors, chanted in front of us, in front of me, the white-haired magician.

We still had to identify three further gigantic aquifers spread across this area, and continue to develop the Lodwar region. Lodwar is the capital of Turkana County and is located 5 kilometres west of the boundaries of the Napuu aquifer, a giant aquifer 700 metres deep, 20 kilometres wide and 120 kilometres long; this water flows into Lake Turkana after running northward from the Kenyan highlands. Before our work, no one had ever detected or drilled for this underground aquifer.

When I first encountered the town of Lodwar in December 2011, it was a dusty and miserable settlement of 12,000 inhabitants, established on the banks of the Turkwel River, whose meagre waters meander through the Rift lowlands between sandbanks and arid volcanic peaks. The river had been dammed upstream and very little water flowed in the winter. Lodwar seemed also to be the capital of poverty, and was the seat of several dozen NGOs and diocesan missions that tried to raise funds to alleviate the suffering of the entire population.

Between 2011 and 2013 I had observed thin women, walking skeletons dressed in rags, sifting through the garbage at the entrance of the city to feed their children. In 2012, I proposed the first boreholes to confirm my deep-water discoveries. This was the first drilling campaign to prove the existence of the Napuu aquifer, located in the Lokichar graben; it was known by this name locally after the Turkana community that had been trying to survive along the banks of the river.

The drillers were sceptical and did not understand our drawings, yet, with the very first well, geysers of water gushed upwards. A second well was drilled, then a third, and finally four wells gushed with water. The aquifer had remained invisible to all the hydrogeologists, despite the proximity of a town that suffered from repeated drought. The water we had found in vast quantities was of excellent quality, so pure it needed no treatment, and it was also renewable. Less than one year later, in 2014, this water was supplying a town rising from its ashes. Since then, Lodwar has experienced no more water shortages, and by 2022 the population had risen to 120,000 persons. And though, to my dismay, swimming pools were built on a few private properties, agricultural prosperity began to grow for 10 kilometres in every direction.

Since that discovery, seventeen villages are supplied by a new water network and 20,000 head of cattle come to drink every day. Small vegetable plots 10 metres square have come to life in the middle of the desert. Women are now growing sorghum, corn, spinach, tomatoes and onions, and almost 1,500 people are now self-sufficient when it comes to their food. Animals that used to wander aimlessly in search of any type of sustenance come to drink regularly from water outlets the Turkana peoples have installed.

Less than one year after our discovery, this providential water had brought dignity and freedom, but, even though I had given everything to the project, the number of wells is still not proportional to the size of the aquifer. The Kenyan government remains mired in inertia and inaction; it prefers to assign the development phase to NGOs, who are better at emergency humanitarian work than at long-term development projects.

Nonetheless, since 2014 there has been a growth in prosperity and the people and their herds no longer die of thirst as they did in 2011. That is, to this day, my sole reward after a solitary itinerary studded with doubt, difficulties and frustration.

But three other deep aquifers still await drilling, in areas where massive drought hits hard: the nomads survive by leading their cattle from one temporary pool to another, regularly abandoning the weakest animals to hyenas. Like all the tribes of the plains, they are also exposed to raids by Merille and Karamojong tribes coming from Uganda and South Sudan, armed to the teeth with automatic rifles.

In 2017, another wave of drought struck Turkana and again hundreds of thousands of cattle died, not to mention the number of human casualties as a consequence of inertia and inaction from the Kenyan decision-makers. At this time I was still busy in Iraq, focusing on restoring some old *qanats* in the mountains of Kurdistan, but faced with this predictable disaster in northern Kenya, I could feel only pain and indignation.

And to our astonishment, the Parisian hydrogeologists from UNESCO contested our discoveries, arguing about the size of the reserves, the renewal of the aquifers; above all, they refuted the results, wasting the immense benefits they could bring to their institution, which had actually paid for this dazzling discovery. Neither I nor the governor of Turkana, who was directly involved with the consequences of our drilling, were invited to Nairobi for the international press conference with members of the government who had gathered to celebrate this amazing success.

Tragically, in 2015, two years after the drilling, the Lotikipi well around which countless women had danced was closed, plugged up as a result of a health recommendation from the Kenyan ministry. The water, they say, is saline and not fit for human consumption. But even in 2022 they had provided no further information and no official figures to back up this claim.

This decision forced the people of Turkana, now totally bewildered, to use a putrid temporary pool of water polluted by the thousands of animals that also use it. Children, on all fours, lap up this water, which will cause them extreme health problems and quite often death. The logic behind this misinformation is baffling.

Almost one year later, the Reuters agency also announced that the water in Lotikipi was salty and fit for neither human consumption nor farming use. The report looked all the more nonsensical after an article in the *Guardian** reported on the outstanding transformation of the area that had taken place after our discoveries. The Kakuma camp, already home to 150,000 refugees, was able to take in 25,000 more in just a few days, at the height of the conflict between South Sudan factions. Other remarkable changes were happening in Lodwar, where new

* "Kenyans frustrated by drip, drip approach to search for water", *Guardian*, Thursday 15 January 2015.

wells were supplying water for the vegetable plots that had been planted everywhere, and which were feeding entire families. The livestock was also drinking from the well water.

Reactions were not long in coming. On 8 March, the Turkana governor stood before the press and sternly criticised the lie,* asserting that Lotikipi water was not salty but free of impurities and potable. He also announced his intention to develop the three further aquifers discovered across Turkana County as a result of my 2013 survey, much like the development in Lodwar.

The discovery had been euphoric; suddenly, I was brought very much down to earth. The mission had been dangerous, urgent and often solitary; driven on by the importance of a successful outcome, it had seemed a wondrous result of all that painstaking work to see the water running out of the holes we drilled into the Earth. And then all of a sudden it wasn't a victory, but rather a resounding defeat as far as UNESCO was concerned. I had become a pariah whom bureaucrats in their air-conditioned cubicles in Paris accused of using off-the-wall techniques to discover irrigable land. They very easily overlooked the many thousands of head of cattle that had died of thirst in 2011, two years before our discovery, and the thousands

* "Turkana official denies reports of unsafe water in County aquifers", *NAIROBI (Xinhua)* 8 March 2015.

"The water in the massive aquifers that were discovered in 2013 in Kenya's drought-ridden north-western county of Turkana is free from impurities," a county official has said.

of people who had similarly died of thirst after great suffering, even though there had been water only metres underfoot. Their response to the Lodwar aquifer discovery and agricultural development was one of silence.

The reality of life and death is constantly on the table in my undertakings, as is the reality of truth and lies, even if it is not always easy to distinguish reality from illusion. And because I work in countries devastated by droughts, I have discovered that the resulting violence does not exist in isolation . . . it is closely related to lies, and makes me remember the Yazidis' sacred belief. Lying and violence pair up, perpetuate each other and lead to destruction in the world. Regardless, there was no way I was going to turn my back on the presence of water that could heal the wounds self-perpetrated by humanity. Water has the power to bring humans back into balance, and to restore peace in conflict zones.

Frustrated and enraged, I nevertheless knew that other discoveries, every bit as significant as this one, were waiting somewhere out there in the world. I had caught a glimpse of the technical and human adventure that lay ahead, and I had also seen the obstacle course that goes with it. The years to come would include fierce confrontations between egos, and political and economic stakes; unbridled appetites for wealth and power would, sadly, continue to frustrate my efforts.

The initial objective was to have brought into production hundreds of wells on the Lotikipi plain, to bring peace through prosperity to this land of conflict. My clearly and often expressed conviction was that a lack of water means war, a lot of water means peace.

The great Lotikipi plain, until now a hostile desert, had been a scene of death during the great droughts of 2011 and 2017. The people in the Turkana region, constantly looking for sources of water, were nomads by necessity, but they had expressed a desire to settle and farm, around wells that did not yet exist. Since then, around Kakuma refugee camp only 40 kilometres to the south of the Lotikipi, vegetable gardens have begun to flourish, protected by thorny hedges; the instinct to settle down comes quickly when there is the possibility of a decent quality of life for both families and the community as a whole. Energy, water, good soil and food security are the four aces of prosperity, security being as vital a component as the other three. It had always been a dream that one day we would be able to provide the first three basic elements to the most underprivileged populations on the planet, to enable them to escape from drought and poverty, and to develop in peace through prosperity.

17: KAWERGOSK REFUGEE CAMP IN KURDISTAN

When I landed in Erbil in the Kurdistan region of northern Iraq in 2015, I found myself once again caught up in turbulent events. Increasing drought impacting the Fertile Crescent from 2009 to 2011 and the effects of climate change had brought about a devastating social revolution in Syria, where millions of people had been displaced through the new order dictated by Daesh.

Kawergosk camp is one of many in the area haltingly built by the UNHCR in August 2013. The devastating progression of the Daesh caliphate in Syria and in north-western Iraq had led to a massive population migration. Kurdistan generously opened its borders to host refugees from Syria and displaced persons from Bakhdida near Mosul and Sahela in the northern Iraqi province of Duhok. I arrived there in January 2015 to evaluate the groundwater potential of this camp, which at the time hosted some 14,000 people.

It was 21 January, at sunset, and the sky was red, streaked with yellow and white. The surrounding plains near Kawergosk camp were strewn with rounded pebbles that very little vegetation was able to push its way through. A hazy range of hills rippled to the west. The temperature was cool, and darkness was about to fall. As the sun went down on the deserted Nineveh Plains in a riot of purple and gold, a faraway humming sound filled the air and long white lines streaked the sky. Bombers were arriving like a score of blackbirds, and their contrails were all converging at the same point on the horizon some ten kilometres ahead of us, in the direction of Bakhdida and Mosul. It seemed we had front-row seats for an anti-ISIS coalition bombardment.

The first explosions reached us in surges. The united forces were piloting their bomber planes over strategic sites in Mosul. After each explosion, the refugees in the camp would stand up and applaud frenetically as if at a football match. Some at the camp had arrived barely three months earlier from the village of Kobani in Syria, chased out by Daesh. They fled with children wounded by exploding mortar shells and were saved from decapitation or sordid slavery by young Kurdish fighters. These young men, seated on sandbags on the garrison house that overlooked the refugee camp, silently observed the bombing, their jaws clenched and muscles tense, ready to fight back.

The fighters were young, between eighteen and twenty-four years old, their faces prematurely aged by combat. They were proud to be protecting their community. One of the young warriors put his rifle down, picked up a mandolin and began strumming a melancholy tune from his childhood. The sunset was taking its time, the bombing had ceased and the planes had veered south; the horizon was dark red and menacing. The young man chanted as he played his mandolin, mourning his comrades lost in war, his gaze empty and faraway. It was cold atop the garrison house and we had to get back to Erbil. It was not a good idea to drive after dark in this region.

In the meantime, the Sinjar area had been assailed and devastated by Daesh; the town of Sinjar, home to nearly 310,000 inhabitants, had been destroyed and entire groups of Shia Turkmen and Yazidis assassinated. Later I was to see their bones filling mass graves scattered on the southern flanks of Mount Sinjar. The European Union recognised the massacre as genocide. Though the actual death count has not been evaluated, thousands of Yazidi men, women and children had been brutally assassinated in just a few days.

The Sinjar district was liberated only one year later.

18: THE MOSUL DAM

During this horrible time of confusion and war against Daesh, I was unable to return to Iraq until 2016, when I undertook a reconnoitring mission between the Mosul Dam and the Yazidis' sacred mountain. This journey had several objectives related to deep aquifer exploration that I had been working on since 2011 with UNESCO throughout Iraqi lands, in view of later reconstruction on a nationwide scale. This ongoing mission had allowed me to take stock of the political and military state of affairs in the zones recently taken back by the peshmerga in northern Iraq between Erbil, the Mosul Dam and Djebel Sinjar, on the Mosul–Rakka line.

More importantly, this ambitious undertaking enabled me to chart the current status of the deep aquifers we had mapped throughout 2015, in order to rebuild these areas devastated by war along the folds of the Zagros range in northern Iraq. This is where the great wheat crops had once existed, subsequently abandoned due to war and mass exodus after the 2014–15 Daesh offensive. I had received an exceptional safe-conduct pass from the governor of Erbil, whom I personally thanked. I was able to cross the defence lines and reach the highly strategic zone of the Mosul Dam. I needed to form my own professional geologist's opinion on the vulnerability of this highly controversial dam, and of the potential impending tragedy.

Another objective of this journey was to meet the top political authorities in Iraqi Kurdistan to know their own challenges and future objectives after the recent opening of Iran, for at that time US president Barack Obama had just signed the lifting of the international sanctions.

On a cold and rainy morning in February 2016, I left Erbil early, my pass safely in hand. My loyal peshmerga bodyguard, Bakhtyar Haddad, kitted out with a bulletproof vest, helmet and Kalashnikov, and with a few bottles of water and two bottles of whisky in the boot of the car, was waiting for me as I left the hotel. The news from the night before was good; both sides of the front were stable all along the Kirkuk–Mosul line.

After an hour and a half zigzagging between checkpoints to avoid being exposed to Daesh gunfire on the shifting, uncertain front, we reached the biggest checkpoint, controlled by the Kurdish Democratic Party, overlooking the Kurdistan Regional Government military camp set up on the left bank of the Tigris, at the other end of the bridge. We stopped to undergo the last formalities, and to pick up information from General Hashim Abdul Baqi Turkistani Nasrulla. This was their last operational headquarters and where all the northern Iraqi operations were centralised. A big map of the Mosul Dam covered an entire wall in the chief of staff's office, showing the positions of the peshmerga opposite Daesh positions still occupying the heights.

The barracks were buzzing with activity. In the courtyard, bottled water and munitions were being loaded into trucks, troops were on the move, armoured vehicles were coming in and others going out. All this movement clearly showed that we were near the combat line, south of the dam near the KAR Group refinery in the Nineveh Plains. We left HQ unescorted. I had to admire the peshmerga optimism: danger was further south, further north, further west; it was "elsewhere", everywhere but immediately in front of them. The Syrian border was supposed to be cleared; we could travel in relative serenity and calm, into the vast plain at the base of the Zagros range. We crossed a dreary plain of abandoned fields, devastated villages and desolate spaces – a war zone, which we passed through in the pouring rain, twitching with anxiety.

The Mosul Dam, some fifty kilometres upstream from Mosul on the Tigris River, supplied most of the region with water and electricity and was vital for the irrigation of wide swathes of crops in the province of Nineveh. The dam, 113 metres high and 3.4 kilometres long, was completed in 1984, a construction that involved nearly 40 million cubic metres of concrete and earth. The BBC reported that the hydroelectric plant could provide up to 1,000 megawatts of electricity. A 2007 American report estimated that the plant had the capacity to supply 675,000 Iraqi households.

To meet ongoing consumption and irrigation needs, the storage reservoir holds 12 billion cubic metres of water, the volume consumed by the city of Paris over the course of ten years. The dam was the pride of former Iraqi president Saddam Hussein. For him, it was representative of the might of his regime, and it bore his name. According to the OECD it was the fourth biggest dam in the Near East.

There was a serious obstacle that raised serious questions: the dam was built on unstable soils composed of gypsum, which dissolves, and limestone, which erodes in contact with water, gradually creating holes in the foundations. As a result, ever since it was commissioned the dam has suffered from structural defects that have earned it the dubious distinction of "Most Dangerous Dam in the World". A report by the US Army Corps of Engineers in 2007 drew attention to the risk of breach of this dam, which has straddled the Tigris River since 1984.

Building the dam on soluble, cavernous gypsum was a major design flaw; deep cavities were burrowing into its foundations. This risk was hugely increased by Daesh's decision in 2016 to blast the dam with landmines. And in any case, it's at risk of breaching at any time due to pressure from the water that builds up after the snow melts in the Taurus Mountains in Turkey. Previously, the Iraqi authorities had tried to consolidate the dam's foundations by injecting pressurised cement into the holes, but Daesh quickly siphoned off millions of dollars of maintenance equipment, tractors, diggers, backhoes and cement mixers, aggravating the problems inherent to the structure. This was the main reason behind the military intervention a few days later, with airstrikes and peshmerga fighters from Iraqi Kurdistan: to rid the Mosul Dam area of any Daesh presence or influence. Any threat to the Mosul Dam and another dam on the Euphrates River, the Haditha Dam, would eradicate the water and electricity supply for the entire country.

Iraqi hydrogeologists had already been sharing their concerns about sabotage of these two main dams: "If these two dams were to go, Iraq would cease to exist," I was told by the Iraqi representative of the Ministry of Water and Irrigation. A simple breach of the Mosul Dam would cause a giant wave 20 metres tall, which in less than an hour would sweep through the city of Mosul 50 kilometres downstream. According to the American embassy in Baghdad, this landlocked tsunami would affect 500,000 to 1.5 million people, who could lose their lives within a few hours if they were not evacuated in time. As for Baghdad, located more than 400 kilometres from the dam, it would still be hit by a 5-metre-high wave; such a disaster would obviously also cause millions of people to be displaced.

A simple precautionary measure would be to evacuate the local population within 5–10 kilometres of the current course of the Tigris, but that was precisely the area still under Daesh control. And Daesh forbade any movement of people; the local population was trapped.

19: CHEMYA! CHEMYA!

After these formalities, General Nasrulla gave us clearance to cross the bridge on the Tigris River, but warned us about the dangers of the proximity of this open space to the Syrian border, which was under the control of peshmerga from Rojava. We were no longer under their protection.

The sun was rising over the Nineveh Plains on this cold foggy morning. We continued our journey, crossing the Tigris River through a dense mist, and drove along the narrow tar road following the Syrian border to the south-west in the direction of Rabia. On the way we could see the abandoned installations of Sufaya oilfield: deserted rigs and torn pipes were scattered across the site, which had been deliberately bombed.

After half an hour we reached the outskirts of Rabia, a village artificially implanted by Saddam Hussein by transferring the Arab population from the Saudi border to the north-east to destabilise Kurdistan. To our horror, we found the village completely destroyed. Most striking amid the debris of flattened homes was the large hospital, which had been hit by British bomber planes hunting out the last bastions of Daesh. Unlike the neighbouring buildings, the hospital looked intact, but its fuel tanks had been decimated, leaving little hope that the building could ever be rehabilitated. The structure was peppered with bullet holes, its interior filled with what looked like a sea of spent shells and the bodies of dead fighters, left for the jackals to consume. The sight of death in this place of healing was horrifying. Abandoned single shoes were sat on top of huge mountains of shell casings, copper wire used to detonate explosives and booby traps poking out of the debris.

The remaining hostile Arabic inhabitants, who by then had been there for twenty years, since the imposed relocation, had collaborated extensively with Daesh. Garbed in dark robes, they kept a low profile and skulked along the walls at a distance from the Kurds who now controlled this village on the border with Syria. We said nothing and kept our eyes down, continuing on our route.

The Syrian border was 500 metres away. It was marked by a long, straight levy studded with dark, dull, enormous fortresses, built in the Saddam Hussein era. This sombre line stood as a separation between immense stretches of abandoned wheat fields; the roofs of the silos were no longer intact, and the grain rotted in the rain. Oil derricks stood immobile as far as the eye could see. In both Iraq and Syria, the production of hydrocarbons had come to a grinding halt with the last bombings, stopping the oil incomes to Daesh and Turkey. We did not go anywhere near the production sites: landmines and other traps were almost certainly still on the ground near each oil production well.

It is hard to comprehend the utter desolation of communities that have been annihilated, and the eerie atmosphere of the destruction left behind. Entire sections of abandoned pipeline littered the place. Villages had been flattened completely, and bent iron framework stuck out of ragged, barely intact concrete blocks from buildings gutted by the bombing. Scarcely three months prior, Daesh had still been solidly anchored in this zone and had gleefully pillaged the oil and gas they then sold by truckloads in Turkey.

After Rabia, heading west, the road to Sinjar was controlled by peshmerga from Rojava, coming straight from eastern Syria.

After a year of deadly combat, the Sinjar district was freed in 2015 – special American forces provided several weeks of support to protect the Yazidi and Shia communities that had fled to the mountains, where they were hunted down like game by Daesh.

We approached the mountain from the north. Near the village of Khadir I noticed that the aquifer I had mapped one year previously was covered by anti-personnel landmines. Coincidence or deliberate sabotage? It was impossible to know, but one thing was clear: the reconstruction of these territories would not be simple.

We entered the sacred mountain of the Yazidis through a dark narrow gorge. The peshmerga who had died in their attempt to get the mountain away from Daesh control were buried where they had fallen. The tombs lay in the shade of a portrait of PKK (Kurdistan Workers' Party) founder Abdullah Öcalan, whom Turkey had sentenced to life imprisonment. Empty white frames stood ready to hold the photographs of each of the men slain in combat a few weeks earlier.

We reached the peak of Mount Sinjar at noon, in the pouring rain, and stopped at the village of Sardasht. We were immediately invited by the village's Yazidi chief into his tent, where a wake was being held. The men were gathered in a dull and solemn silence, their faces portraits of despair; the community chief, I learned, had suffered the extermination of his entire family. His wives, daughters and granddaughters had been kidnapped and sold into slavery in Syria and Iraq. The older female members of the family, deemed undesirable and therefore unsellable, were not kidnapped but instead assassinated and dumped into the mass graves on the nearby plains, along with the male members of the family. These venerable elders, mourning one more loss, were the sole survivors of a community decimated by Islamist fighters from Syria, Arabia, Morocco, France, Afghanistan and even Chechnya, brought in by the truckload, imposing their law of rape and plundering and violent fury on these poor peace-loving communities.

Our hosts offered to share their meal with us. Sheikh Zakho was the spiritual leader of the Yazidis, all of whom were pacific and industrious farmers. His face was lined with deep sorrow; he and the others were haunted by the ongoing violence and genocide. The South Sinjar plain, 800 metres from the front line, was still under Daesh control. Peace and security were not yet established. The northern side of Mount Sinjar was currently being cleared of mines, all the way up to one of the strategic aquifers scheduled for drilling. The weather outside seemed to capture the atrocities we were witnessing. Cold, dense fog engulfed us, and we were unable to see the terraced fields on the sides of the mountain, where Yazidi farmers had eked out an existence before they joined the guerrillas against Daesh, or were lost for ever in mass graves.

After driving up the hairpin turns of Mount Sinjar in a thick cloud, we started down the southern side in the direction of the town of Sinjar, or what remained of it. It was an apocalyptic sight; the entire town was burnt out, gutted, a heap of twisted iron and blackened concrete. There was not a single building still standing. As I stood in the midst of the destruction, I felt all the more determined to stand up to the jihadist movement, for the scenes were reminiscent of those from the Bataclan in Paris. All of a sudden, the future of Europe seemed uncertain, and standing up to this scourge even more vital.

Ruins of walls were covered with Islamic State graffiti explicitly referring to the ethnic and religious origin of the slaughtered Yazidi, Shia and Sunni inhabitants. The Islamic graffiti writers had even dared to invoke the justice of their prophet Mohammed, and signed their name to their crime, sometimes even in Russian for the Chechen Daesh members who had come to steal the modest belongings of the Sinjar farmers.

Leaving the demolished town behind, we took a road full of holes to the front line, known as Camp Dumez, to meet Lieutenant-Colonel Khalid Hadj Hassan, who was very surprised to see foreigners so close to the scene of battle. Camp Dumez had been given its name by the French company subcontracted by Saddam Hussein to build new infrastructure in

Kurdistan. It was positioned on the strategic east–west road from Iraq to Syria, south of Mount Sinjar, and was cut in two by the front line. We went to the outpost known as the "Bastion front line", which controlled the road leading to the Arab village of Ranbusi, occupied by Daesh. The defence line was nothing more than an earthen embankment bulldozed to chest height and topped with sandbags. It stretched several hundred kilometres from the Syrian border to Mosul, a fragile fortification with arrow gun-slits for soldiers to shoot through every 200 metres. Placed periodically along the fortifications were reinforcements in the form of 40-mm-thin mortar or a vehicle armed with heavy machine guns. These were World War Two Russian DShK (pronounced "Douchka"), which could easily be repaired by the peshmerga with a hammer and an anvil. The rest of their equipment consisted of spoils from the enemy, which they took at every opportunity.

Bullet casings, still warm, were strewn everywhere. A gunfight had just taken place and the feverish fighters were silently smoking cigarettes. All of them were brothers-in-arms, from the same village – uncles, cousins, nephews, bound by blood to their land. This defence line was their last refuge, and they had taken it back only a few months before. Peshmerga fighters paid a heavy toll. Each one was a survivor; each had lost kin. Each had mothers and sisters captured from neighbouring villages in Iraq, sold into slavery to Arab traders in various marketplaces. Those women considered most attractive were traded in Syria, sold to the Daesh big shots in the markets in Raqqa and Deir ez-Zor, the fief of al-Baghdadi's Islamic State. The peshmerga soldiers hid their bitter sorrow and rage. Nothing would ever be the same; the consequences of these crimes would long influence the future of the entire region and its relationship with its Arab neighbours.

Lieutenant-Colonel Benguin explained the situation in perfect English: "The tragedy is that we Kurds are not considered to be a nation, and so friendly countries such as France, Germany and America cannot sell us weapons. The weapons you see here were sold by the Americans to the Iraqi government and were seized by Daesh during the battle for Mosul. We captured these arms from the enemy. We took what was still in decent shape after the fighting, during our last battles.

"We have just one MILAN light anti-tank infantry missile, unofficially supplied by the Germans, and forty assault rifles to defend ourselves with on an almost 100-kilometre-long front. We used this missile from a distance of 800 metres to destroy a truck packed with explosives that was bearing straight down on us at the front line, to breach the line. Our enemy loves death. We love life," said the colonel. "That is the real difference between them and us."

At around 5 p.m., when the wind changed direction, Daesh artillery fired two shells that fizzled on landing. No explosion. The soldiers inspected them with an experienced eye, and then reacted immediately to two bursts of grey-brown smoke 800 metres from where we stood. "Chemya! Chemya!" Chemical explosions make very little noise; if the eagle-eyed soldiers hadn't shouted a warning, we would not have noticed – they certainly saved our lives. Everyone took out their gas masks, none more sophisticated than the surgical masks so prevalent during the pandemic. Europe was banned from delivering gas masks to the Kurds, for gas masks are considered war equipment, as are bulletproof vests. These men, seemingly abandoned by the world, had nothing but their courage with which to protect themselves.

Their only real recourse, if attacked, was to radio the coalition bombers that flew continually over the combat zone in a constant, faraway hum, cleaning up the land below. Air support could take one to two hours to arrive. The Kurdish soldiers especially appreciated the prowess of the French pilots, who now flew with even greater determination since the terrorist attacks at the Bataclan in Paris.

Lieutenant-Colonel Benguin resumed his mournful tale: "The peshmerga directly defend the freedom of the Europeans. What you have here is the last front line against advancing Islamist fundamentalism, and we are fighting with the most paltry of equipment to keep fundamentalism at bay and Daesh from advancing, to keep threatened populations from fleeing to Turkey and from there to Europe." Drawing in a deep breath, he paused for a moment, looking worriedly at the movement of the two grey-brown clouds of gas that had emerged a few hundred yards away by Mount Sinjar. "We are the last rampart against the crimes and genocides that seem to leave Europe completely indifferent. I'm not even referring here to the United Nations, who complain that we hand over foreign, in particular French Islamist, war criminals captured with their weapons in hand, to the courts in Baghdad. Europe worries us. Europe prefers to speak with the Turks while criticising Bashar al-Assad. Europe worries us as much as the Turks who shoot us in the back and take in wounded Daesh fighters and rearm them to fight us. Bashar Al-Assad doesn't send terrorists to kill us in Kurdistan, doesn't send terrorists in Europe, so when will Europe understand who her true allies are?"

Gas from the two chemical shells was spreading across the plain, beginning to irritate our skin and lungs. We had to leave as quickly as possible for the ruins of Sinjar, leaving the brave Kurdish soldiers on the front line.

20: A NIGHT WITH THE SURVIVORS

We were to spend the night in the Yazidi forces' HQ. This was a stone building that was originally the Daesh HQ; they had had no time to blow it up when they were forced out. We rattled down a muddy path as darkness fell, rain cascading down and drenching the plain. It was very cold and wet; the base of Mount Sinjar had disappeared in the thick fog. We were only 800 metres from the front line, which separated us from the Nineveh Plains occupied by the enemy on the Raqqa–Mosul line.

The general of the Yazidi forces was waiting for us at the entrance to his house and welcomed us warmly. We proceeded to a beautifully – and bountifully – laid table. How had they procured such lovely cheeses, fruits and vegetables, mineral water and raki, to celebrate our arrival? My bodyguard Haddad, who had kept this surprise from me, seemed delighted. We were in friendly territory and that night our protection was guaranteed by the nearest outposts, whose radios could be heard going back and forth with the various units deployed along the front.

The night fog continued to make me nervous, however, for visibility around the camp was no greater than 100 metres. Bomber planes flew relentlessly over the combat zones, jerking and jolting in the wind and rain. Winter had taken over, and it brought with it a thick coat of sadness. The Yazidi general, a venerable patriarch, gave us the only available bedroom; it had been that of his daughter, who had vanished more than a year before – her location was still unknown. The options were hard even to contemplate – she might be a slave in a house in Raqqa or Mosul; she might be buried in one of the mass graves on the southern flank of the mountain, or perhaps she was dead by some other means. No one could lighten the weight of his grief, and despair hung heavy in the air. His wife, too, was missing – both had disappeared in the upheaval of a year earlier. In a low, quiet voice, he told me what had happened.

"It was 3 August 2014, a very hot day. That morning, my wife and my daughter had just started out on foot very early for Sinjar market when hordes of Daesh fighters appeared in the plain, coming from Syria, from the south. We knew that danger was imminent, that something could happen, but we had no idea it would happen at that speed. At the time I was in the fields on my tractor. Danger arrived from the south-west, in a flash, in a cloud of dust like a flight of locusts. Horns were blowing and guns firing. I immediately knew we were in trouble. I jumped on my tractor, went straight into town to gather the whole family. The marketplace was full of panic and shouting; I couldn't see my family. I thought my wife and daughter had escaped on the long, winding road up Mount Sinjar, their ultimate refuge. We had discussed all this ahead of time, had even rehearsed it in advance, although not at this speed.

"This steep road to Mount Sinjar was already full of panicked movement: trucks, tractors, cars, all piled high with hastily filled suitcases, women and children perched on top of bags, in an overwhelming din of horns, cries and screaming. It was a scene of apocalyptic, end-of-the-world panic. I couldn't move an inch, never mind attempt to find my family, in this chaos. I left the road and went straight up the rugged mountainside on my tractor, hoping to get ahead of the panic and rejoin the road in front of the crowd. I tipped over on the edge of a sharp rock. I was thrown out of the tractor, which overturned and rolled down the slope." He paused for a moment, his gaze empty and sad. It was hard for him to talk but he made the effort. Haddad encouraged him to continue.

"I lost consciousness and was left for dead on the ground. I came to in the middle of the night, my face covered in blood. I could see nothing and thought I was blinded. My forehead was one big gash, but I was able to move. I'd lost a lot of blood and was desperately thirsty. A terrible smell burned my throat. Slowly I lifted myself off the ground, and the pale glow of the moon revealed the tragedy that had occurred all around me. The smoke and glowing heat of burnt-out vehicles wound its way all along the entire mountain road, up to the very top. There was not a single sound, not a moan or a cry. When I got to the road, I saw by the light of burning cars piles of clothes along the roadside – suitcases and bags ripped open and scattered, girls' shoes and underclothing, children's shoes." His breathing was ragged; the nightmare, the guilt, was clearly still fresh in his mind.

"It has been one year already, but it is as if it was yesterday. Scenes you wish you could forget remain with you night and day, like poison that flows in your veins, poison that brings back unbearable, throbbing pain that leaves you helpless. The truth quickly became clear, although at first I refused to believe it: the end of the world never applies to oneself. Then, too thirsty to think, I headed to a dry stream and found a puddle, where I lapped up water like an animal. Indeed, I had become a hunted animal, with no personality, emptied of life, with no more courage or identity. Daesh had taken everything from me; perhaps Daesh was still waiting to finish the job, and in my despair I hoped it was so. But drinking that water brought me back to reality. The faraway whirring of a helicopter gave me new courage: Daesh has no helicopters and I knew that meant the Americans had arrived a few days earlier from Baghdad. That gave me the strength to finish climbing up the steep mountain. I arrived in Sardasht the following morning, at dawn, and there I saw the survivors. They told me what had befallen them and I remembered it all up to the point I fell from my tractor. Then they told me the rest.

"The fleeing vehicles were caught up in one immense traffic jam. Total anarchy reigned on the road. Daesh attacked the escaping inhabitants and trapped them family by family along the road; they set the vehicles on fire with petrol and grenades. They separated men from women, opened all the suitcases and bags, stole all the jewellery, money and mobile phones. All men and children over ten years of age were rounded up, taken by truck and methodically shot below the town of Sinjar. The youngest women were separated from the others and taken by bus into Sinjar."

He later learned that these buses then took the young women to Mosul, where they were locked in a large house with hundreds of other young women. Daesh fighters would come to the house and select three or four girls at a time and take them to their lodgings. All the women were routinely beaten and raped, victims of sexual slavery. They were traded like cattle to markets in Mosul and Raqqa, and price tags ranging from $5,000 to $50,000 were placed around their necks. More than 2,500 Yazidi women disappeared that day, and over 4,500 men were assassinated by gunshot at point-blank range. The older women were less fortunate than the men. They were piled into a freshly bulldozed trench and buried alive. Others were drowned in a fishpond just next to Camp Dumez. Any survivors who tried to flee were gunned down.

A list was drawn up by Narin Shiekh Shamo, a Yazidi activist now living in Iraqi Kurdistan, of the names of the 4,601 missing Yazidi women. Kurdish authorities state that they have saved approximately a hundred Yazidi women from the clutches of Daesh, sometimes paying a ransom through the intermediary of Arab tribes transplanted there during the Saddam era. All the survivors spoke of systematic rape. In the Daesh propaganda magazine *Dabiq,* Daesh acknowledged the practice of giving Yazidi women and girls to its fighters as "war booty". Daesh even sought to justify its use of sexual violence by stating that Islam authorises men to have sexual relations with, and to sell, non-Muslim "slaves", including barely pubescent young girls. Human Rights Watch stated that these declarations are further proof of a widespread practice and a systematic course of action implemented by Daesh.

Though the American army began humanitarian operations after the Battle of Sinjar, it was only in late November 2014 that President Obama, after incomprehensible hesitation, finally sent ground forces, including special commandos, to back up the peshmerga. It was very difficult to fall asleep that night in Sinjar, in the bedroom of the young Yazidi girl who is most probably for ever lost to what remains of her family.

All night long the relentless drone of planes reminded us that we were in a combat zone. The crate of hand grenades given to me by the Yazidi general in the event of attack that night only served to reinforce my unease. The front line was close by, the bad weather conducive to attack. Each of us had to be responsible for our own safety – simply being there made us fighters; there was no other choice, at least for this night.

It took a long time to fall asleep as I listened intently to all the noises of the night through the rain pattering on the sheet-metal roof. I mentally reviewed all the possible scenarios that could unfold in the event of attack, and I couldn't help wondering what our chances of survival would be with a few grenades in my inexperienced hands . . . My thoughts then turned to the wheat fields, the war-torn silos and the deep water that would one day be accessible when peace was re-established. I felt a surge of new hope for a country that needed to be entirely rebuilt. But would that even be possible? A previous experience in Rwanda several years earlier had left me doubtful; I had witnessed the trauma of survivors attempting to live normally again, cheek by jowl with neighbours who had assassinated their families. Deep down I was assailed with disgust and sorrow; it was hard to balance it with optimism and hope.

21: LOST IN KURDISTAN

My unexpectedly emotional journey to the Yazidis' sacred mountain focused initially on the deep aquifers necessary for post-conflict reconstruction. But this trip was the start of more general thinking, furthered by an intense exchange with Kurdish military staff stationed along the Tigris near the Mosul Dam, and with Kurdish politicians in Erbil and Sulaymaniyah. In 2016 the situation with the Iraqi Ground Forces was still quite unstable.

Although Kurdish forces had in 2014 wrested control of the Mosul Dam from the Islamic State jihadists who had won it ten days earlier, they were not yet back in control of Mosul. Control of the dam was a key issue for water and energy distribution in northern Iraq. After leaving Sinjar I met the new governor of Erbil for a debriefing. He told me his view on the current situation in Iraqi Kurdistan. "The Kurds' situation is more complex than it appears," he said, painting the scene with a measured moderation. "In this conflict, what is the real importance of Daesh? It is clear that Daesh is going to have to get out of Iraq and obviously out of Mosul. That's going to take time and requires the right approach. Attacking Mosul head on would lead to a human tragedy and loss of materiel like in Fallujah or Ramadi, which became two utter wastelands. The people of Mosul are well educated, but they are also mistreated and starving. The people themselves have to rebel. In this context, the Islamic State has no future. The only ones who can still help Daesh in Mosul are Turkmen, who don't have much education. Mosul is a city with 2 million inhabitants."

Our discussion continued later at the Hotel Divan in Erbil, where I also spoke with the Kurdish general Saqar Sulivani, from the Iraqi staff. He is a Ba'athist, head of the Sunni secret service with the central government in Baghdad.

For him, the Kurdistan regional government represents the western values that are needed to advance the country: secularism, democracy and equality for women. He was highly critical of President Obama, who he said blindly followed instructions from the Iraqi central government. A blind eye had also been turned to former prime minister Maliki, who expropriated $41 billion over eight years for the Shia Muslims. Today, Iran-backed Shia militia in Iraq is better equipped than the Iraqi army itself.

The crisis situation in 2016 had been perpetuated by a number of different factors: a drop in oil prices, the sudden influx of 2 million refugees (670,000 of them displaced people living in camps in Kurdistan, whose own population is only 5 million), and Iran's adoption of Russia's agenda in Syria without thinking about how the Kurds would fit into it in the future. At the time of our discussions, the thin red line that established the divide between Kurdish forces and Islamic State jihadists passed south of Mount Sinjar. Daesh controlled both the biggest deep aquifers in the Kurdistan foothills, and the main oilfields in northern and central Iraq. The coalition forces would have to push back this front at least 100 kilometres to the south-west, past the useful area of the oilfields and the aquifers.

Of course, the constantly rising number of displaced people was a major and ever-growing concern. One and a half million Yazidi, Christian, Chaldean and Zoroastrian refugees had been driven away from the Nineveh Plains and had taken refuge in Iraqi Kurdistan; an extremely small country was suddenly absorbing these refugees, in turn stemming emigration to Europe, from which Kurdistan received no aid. The population of the city of Erbil doubled in size, creating many water-related issues that will remain unresolved unless deep aquifers are strategically developed in the near future.

It is encouraging that Iraqi Kurdistan has exceptional vitality. The country, which was bombed to smithereens and routinely threatened with genocide organised by Saddam Hussein, in 1991 took a great economic leap, with positive impact at the regional level. Kurdish firms established in Turkey began to prosper with Iraqi-Kurdish oil money. Construction work made enormous progress with the reconstruction of the outlying urban areas around Erbil. Iranian-Kurdish businesses such as import–export of consumer goods, fruits and vegetables and hydrocarbons, took their share of the market. After my meetings with Kurdish leaders, I began to see the future of the region more clearly, with the creation of the energy corridor between Syria, Iran and Russia, thanks to stronger ties with Assad in Syria and help from Russia.

The Sykes–Picot agreements* of 1916 created territorial divisions that had an overwhelming impact: they arbitrarily determined the shape of countries, their economic potentials and their access to the sea. These changes should now drive the Kurds to play a historic hand towards an alliance with Iran, Syria and Russia, without harming the Turks.

* The Sykes–Picot agreements were secret agreements signed on 16 May 1916, after negotiations between France and Great Britain from November 1915 to March 1916 on the partitioning of the Near East into several zones of influence. This was in essence the dismemberment of the Ottoman Empire.

22: MAJOR CHALLENGES LEAD TO MAJOR OPPORTUNITIES

We already know the consequences of climate change: one-fifth of the world's population and one-third of the Earth's surface is threatened by the impact of desertification. This is particularly true in Asia, in the Sahel in Africa, in North America and around the Mediterranean.

The areas I have worked in for these past few years are just the tip of the iceberg, because what lies further down the road due to climate change is a veritable deluge of water-stressed areas. Developing countries are no longer the only areas threatened, as we may have thought in the past. Extreme climate events have been affecting every country indiscriminately for several years now.

It is hard to know exactly what predominates among the multiple causes of climate change:

– It could be the natural geo-climatic cycle, which has occurred repeatedly for millions of years.

– It is certainly exacerbated by the burning of fossil fuels, which in turn increases carbonic gas and methane in the atmosphere.

– It could even be due to the cycle change caused by solar flares.

– We have proof now recorded within geological layers that major volcanic eruptions have impacted and even destroyed civilisations, such as Thera, now known as Santorini in the Greek Cyclades, which caused the end of the Minoan civilisation.

– Plate tectonics began to play a major role in climate regulation and biologic evolution on Earth half a billion years ago, and the last plate tectonic event generated the four major glaciations and deglaciation cycles with the creation of the Costa Rica–Panama land bridge, which separated the Pacific Ocean from the Atlantic. This land bridge created a major oceanic circulation change with the rise of the Gulf Stream which started to generate massive snowfall onto the North Pole and Greenland. This corresponds to the Ice Age, as has been recently discovered.

The reality is probably much more complex, and explaining natural phenomena of this magnitude requires complex equations and detailed description backed by strong scientific evidence.

– But ultimately, the demographic growth and the concentration of populations in urban settings remains the major crisis driver.

– Demographic pressure and industrial expansion has induced intense deforestation for fuel and charcoal for domestic use and contributes to extreme weather contrasts, higher temperatures, droughts, floods, huge wildfires, storms, rising sea levels, soil degradation and acidification of the oceans. These events are rapidly intensifying, threatening infrastructure, health, water and food security.

The ecological consequences can be huge and largely unpredictable; and, of course, along with demographic concentrations, we've seen a procession of recurrent pandemics such as a worldwide spread of SARS, H1N1 (commonly known as swine flu), Ebola and, most recently, Covid-19.

We can already predict that population increase combined with the speed of international exchanges of goods and persons is ushering in an era of massive pandemics.

If we cannot fight against nature, we can work with it by adjusting our sails to the winds and extract the best opportunities available to us. Any attempt to change the course of climatic events is, from the standpoints of political, technical and financial feasibility, purely a pipe dream.

Today, we have a pattern of available scientific solutions to help humanity to choose the right paths, compared to our ancestors in the Stone Age who faced several extreme climatic disruptions and drastic diseases which led some groups to the brink of extinction: the international vaccination process, for example, is becoming extremely responsive in terms of speed and efficiency, as has been demonstrated in the case of Covid-19.

For the last two centuries, science has led humanity to better understand that Mother Nature has a wealth of biological treasure to offer, if respected and observed closely. In 2008, the Amazon Research Institute demonstrated that many volatile molecular compounds are exchanged between plants, and all transmit an identical chemical message with an active vocabulary. This message is naturally emitted by plants to communicate the existence of a potential threat to other nearby plants, so that they can protect themselves accordingly. The institute has demonstrated that the methyl jasmonate molecule conveys a universal message of distress, in all tropical trees in particular, of the need for water in the case of severe drought. The molecules tend to aggregate the cloud moisture in the form of raindrops, which initiate the rains. So, the rainforest is intrinsically linked to rainfall or permanent mist.

In addition, we are learning more and more about plant neurology. Some plants work together with certain types of bacteria that have the power to "fix" atmospheric nitrogen. This is the case with vegetables that have learned to partner with these bacteria by attracting them to their roots to offer them the sugars necessary for their growth. Thus, the nitrogen that makes up more than 70 per cent of the Earth's atmosphere is made available free of charge to bacteria that produce ammonium to fertilise sterile soils, without the need of human intervention with artificial fertilisers.

Given the intensive deforestation of many countries and the decline of surface water resources, we can see how the predictable cycle of severe droughts followed by famine leads to civil wars. That is why the emphasis must be on public awareness, education, and investment in intensive reforestation. Tree planting must be planned wherever possible, including vegetation on the roofs of buildings.

China, for example, faces the very serious problem of desertification, with 2.5 million square kilometres of land under threat. This process has triggered a major opportunity: in 1978, China launched its "Great Green Wall", designed to replant 90 million acres of new forest in a band stretching 2,800 miles across the north of the country. This action to reverse decades of desertification is the right choice: it will take time and vast effort, but will eventually be successful if a broad variety of endemic tree species are introduced in the process. In Davos in May 2022, China announced its goal to plant and conserve 70 billion trees by 2030 as part of the global tree movement. And now Africa is launching its own Great Green Wall. Hopefully we're proceeding on the right track.

23: ULTIMATE REWARDS

I will never forget the moment, early one morning, when I heard the shrill ring of my mobile in the cold fog of the Kurdish mountains of Sinjar, northern Iraq. It was the day after we had been gassed by ISIS shells along the front line of Camp Dumez. It was my friend Joseph Patrich, Professor of Archaeology at the Hebrew University of Jerusalem. I had once worked with him using radar tools to search for the entrance to the tomb of Herod on the hillsides of Herodium in Israel, beneath the great fortified hill Herod had constructed for his burial place.

He was calling with surprising news, which he had probably learned from Mossad: the Space Foundation in Colorado Springs in the USA had just inducted me into their Hall of Fame for my work on deep water in Africa and the Middle East. My invention of the WATEX™ system had also been awarded a prize.

I was unable to find the right words to express my emotion. On that particular morning, after our experience the day before and an almost sleepless night, other concerns were at the forefront of my mind. There was a sensation of joy, pride in my hard-won and long-fought-for recognition, but with it came a sense of shame. I felt unworthy of any kind of award or recognition in the context of such tragedy.

*

On 14 April 2016, a few weeks after the high-risk expedition in Sinjar, I was standing, incredulous, in the luxurious lobby at the Cheyenne Mountain Space Force Station, the American military base in Colorado Springs. The award ceremony took place with much pomp and deference, and in the presence of 1,200 senior space leaders heavily protected by secret service armed forces. My presentation panel was composed of the astronaut Leroy Chiao, Gwynne Shotwell from SpaceX, Dan Lockney from NASA, and my friend and hero, a man who landed on the moon, Buzz Aldrin. The guests applauded when the panel announced that, thanks to NASA space technology, I had been able to develop the WATEX™ system to significantly improve the lives of millions of people in some of the most dangerous places on Earth. It was a proud moment for my wife, and for my friend Dr Saud Amer, who also received an award for having supported our development programmes through all these years since the Darfur crisis in 2004.

I vividly remember the emotion I felt when Buzz took me by the shoulder to congratulate me for finding water on Earth, and then pressed me to find water on the moon! Why? I was stunned by his request. Because, he said, whoever controls the moon will save our planet. I was reminded of the famous quote from General de Gaulle: whoever controls the upstream controls the downstream.

Now our upstream was the moon and I was confronted with the ultimate space challenge: whoever controls water resources and energy on the moon also controls our future on Earth!

It took me some time to realise how true this affirmation was. With access to water and unlimited energy on the moon, thanks to tritium, known as "regolith", stored in the moon's sediment, it would be possible to produce hydrogen and oxygen, the two basic elements necessary to produce life. The energy produced could propel new rockets to repair telecommunication satellites orbiting the Earth that had been damaged by solar eruptions. Without telecommunications, humans would return very quickly to the Stone Age.

Anticipation is the key to our survival, and the stakes now lie in space; but, as on Earth, with water.

24: WHAT CAN WE HOPE FOR?

"The only thing necessary for the triumph
of evil is for good men to do nothing."
(source unknown, although quoted by Sir Winston Churchill
and often attributed to Edmund Burke)

For a geologist, climate change is not a new phenomenon. Changes have been occurring for millions of years, mainly caused by volcanic eruptions, solar flares and, fortunately more rarely, by meteorite impacts. Events such as these punctuate the emergence and disappearance – or the evolution – of species.

After the enormous asteroid or comet struck the Yucatán Peninsula 65 million years ago and within a few hours extinguished almost 80 per cent of all species on Earth, mammals filled the gap left by the dinosaurs and evolved into many sophisticated species that we would recognise today – including humans! Most of the plant and animal groups from the Cenozoic era have been evolving until now, and continue to evolve and change.

Volcanic eruptions and solar storms are far less dangerous than asteroids, but they still affect the climatic cycles that have successively impacted both the prosperity and decline of the great civilisations, from Mesopotamia to Egypt and later the Roman Empire. We all know about the disappearance of Pompeii in AD 79, but the sudden disappearance of the Minoan civilisation, 3,200 years ago, is less well known because it is older. It was caused by the massive volcanic eruption of Thera in the Cycladic islands. This event effectively closed the Bronze Age, with the fall of the Egyptian Empire of Ramses II, followed by the fall of the Mycenaean civilisation in Crete.

Thanks to archaeologists and palaeontologists working on different timescales, we are now beginning to understand what really happened. And there is no reason why our contemporary civilisations should entirely escape these laws of the universe.

We know that with the Industrial Revolution just over a century and a half ago, along with unprecedented population growth, an aggravating and entirely new phenomenon emerged. The positive effects of the Industrial Revolution are well known: enhanced quality of life, with momentous improvements in science and healthcare and an increase in the longevity of populations, alongside the development of new means of communication and transport, extraordinary agricultural productivity and the emergence of an efficient food supply chain.

Thanks to scientific advances in fields as diverse as biology, medicine, physics, chemistry, genetics and many other sectors, most of humanity no longer knows the famine or infant mortality of our grandparents' generation. All this progress has enabled dizzying population growth unparalleled in the history of humanity.

But the climatic upheavals that we are witnessing today reflect the negative effects generated by this industrial progress over the past half a century. They are beginning to manifest themselves on a global scale. We have seen how progressive deforestation due to the extension of agriculture and factory farming along with uncontrolled urban growth lead to the pollution of the water in our rivers, lakes and oceans. With the acceleration of deforestation since the Industrial Revolution, monoculture reforestation (with non-endemic, fast-growing trees transplanted from a different environment) has become very vulnerable to climatic changes and can ignite in giant fires. All these phenomena cumulatively lead to the rapid increase of extreme thermal differences, which, on a planetary level, unleash tornadoes and typhoons at a frantic pace. They also result in extreme droughts, followed immediately and in the same areas by extreme floods. Extreme droughts trigger dust tornadoes and gigantic wildfires in all latitudes and in both the northern and the southern hemispheres. Extreme rainfall destroys infrastructure and crops, producing gullies in land stripped by fires of the protection of vegetation, leaving a sterile soil like the skin of a burn victim.

We should never forget that humanity is a part of a whole that we call the universe, a part limited by time and space; and to quote Einstein's 1950 speech on the deathbed of his friend, Rabbi Robert Marcus, "Our task should be to free ourselves from this prison by expanding our circle of compassion, in such a way as to include all living creatures and all nature in its beauty."

I am not predicting the end of the world – I am simply describing the end of *a* world. If our generation is observing the end of a cycle, it will also contribute to the emergence of a new cycle.

Thinking back to my founding of my company, I could not have imagined then where my inventions would have taken me twenty-three years later. The complex algorithm now used in the search for deep water resources was born first from a great vision for humanity. It was created by agglomerating fragments of my personal experiences with available data: oil exploration, geology, geophysics, image processing and mathematics, elements of quantum mechanics, and declassified military satellite images from the Strategic Defense Initiative during the Reagan era.

But that was too much theory and not enough practice: I needed a practical field of application to prove the accuracy and effectiveness of this algorithm.

The Darfur crisis in 2004 offered me this opportunity. Within a few months, the immense effectiveness of this invention was confirmed when I saw water gushing from the desert.

My dream of saving several million people in a few months, in hundreds of United Nations refugee camps between Chad and Sudan, proved possible and was beginning to become a reality. Since that extraordinary time, I have mapped entire countries, such as Iraq, Kenya, Somalia, Costa Rica, Mauritania and Niger.

My induction into NASA's Hall of Fame was followed by the French military order of merit, the Légion d'Honneur. Since then, at the request of the United Nations, I have not stopped travelling the world to contribute to the reconstruction of conflict zones.

My name is Dr Alain Gachet, but I am no physician; my purpose is to find water to heal the wounds that man has inflicted upon himself and his environment.

In Iraq, just after the fall of Mosul, I was engulfed by overwhelming grief. My bold and brave companion and friend, Bakhtyar Haddad, had been killed on the Mosul front in June 2017. I was told he did not suffer, that he died instantly, alongside two French journalists, ambushed by an explosive device that was set off remotely. Some consolation; but still, the loss of a friend and colleague in the fight for water remains almost unbearable.

Earlier, in northern Kenya, my guide, translator and friend Daniel Lochomin, with whom I had dreamed of building a farm near a well in Kachoda, north of Lake Rudolf in Turkana, was assassinated as he tried to defuse intertribal conflict.

May the All-Powerful celebrate the memory of the righteous water warriors.

ACKNOWLEDGEMENTS

To my wife Frederique Gachet, a talented journalist and writer, who has been encouraging and supporting all my efforts (and has sometime endured my lies as to where I am travelling to, so as not to panic her . . .) for more than thirty-five years, healing the pain left behind from working in some of the world's most dangerous places.

My deep and personal acknowledgements go to my dear friend Catherine Tate, who took care of the first translations of the initial French text. Her passion for my work has been inspiring.

Helen Frances, recommended by my first Arcadia Books publisher Piers Russell-Cobb, has also contributed to the articulation of the complex stories that illustrate several aspects of climate change.

Without forgetting my friend Katharina Bielenberg, my new publisher at Arcadia Books, now a part of Hachette, who has taken my manuscript and encouraged me as I put the final touches to the book.

ABOUT THE AUTHOR

Alain Gachet, CEO and founder of RTI Exploration, born in Madagascar in 1951, is a nuclear physicist specialising in quantum mechanics, a mining engineer, petroleum geologist and geophysicist. He has worked as an international expert for the United Nations and the USGS on the Science for Diplomacy programme.

After working for twenty years in oil exploration, Alain Gachet made a major change in shifting from oil to water. In 1999 he created RTI (Radar Technologies International) to explore deep aquifers all over the world, using a secret algorithm based on quantum mechanics, geology and geophysics.

He now seeks deep potable water in war-torn areas in Africa and the Middle East and is responsible for several major aquifer discoveries in Iraq, Sudan, Chad, Niger, Somalia, Kenya, Ethiopia, South Africa, and more recently in Costa Rica and Chile for the purpose of reforestation.

In December 2014, Alain Gachet was appointed Chevalier de la Légion d'Honneur. In 2016, recommended by NASA, he was inducted into the Space Technology Hall of Fame, a highly selective club of scientific humanists, at the Space Foundation in Colorado Springs, USA, for having significantly improved the lives of millions of people around the world using fundamental physics and space technologies.